T0336096

Front End Engineering Design of Oil and Gas Projects: Critical Factors for Project Success

Quite a large number of major oil and gas projects are failures with respect to their costs, schedules, and operational performance. Owner companies and contractors are struggling with the issues causing these failures. The Front End Engineering Design (FEED) has been identified as an important factor that plays a key role in determining the success of a project. However, the FEED and the associated Front End Loading (FEL) do not get the attention they deserve from the players in the business, namely, the owner companies, FEED, and EPC contractors. While academic studies on the FEL and its failures are available, how the seeds of failures are sown during an actual project FEED remains a mystery. The details are usually buried in the rubbished computers and hundreds of files that are shelved in companies' offices.

In this unique book, two experienced professionals, one from an owner company and the other from an international EPC contractor, whose interests often oppose each other, join to give their perspectives about the project lifecycle, its governance structure, gate system, complexities, contract models, and quality measurements.

In the second part of this book, they present case studies of projects gone wrong, due to mismatches, errors, and inconsistencies in the FEED. These case histories reveal how avoidable gaps and errors creep into FEED resulting in project failures and how the review systems fail to detect them. Technical and business professionals seem to underestimate the importance of FEED in capital-intensive major projects, while focusing on short-term goals. The underlying causal factors need to be addressed and resolved in time properly, for ensuring success of major oil and gas projects.

Written in a concise and practical style, with key takeaways at the end of each chapter, this book will be a useful guide for practicing project and engineering professionals in the oil and gas industry. Senior students and researchers will find ideas and viewpoints given in this book worth exploring further.

Front End Engineering Design of Oil and Gas Projects: Critical Factors for Project Success

Perspectives, Case Studies, and Lessons

G. Unnikrishnan and V. Pratapkumar

CRC Press
Taylor & Francis Group
Boca Raton London New York

CRC Press is an imprint of the
Taylor & Francis Group, an **informa** business

Designed cover image: © G. Unnikrishnan and V. Pratapkumar

First edition published 2024
by CRC Press
2385 NW Executive Center Drive, Suite 320, Boca Raton FL 33431

and by CRC Press
4 Park Square, Milton Park, Abingdon, Oxon, OX14 4RN

CRC Press is an imprint of Taylor & Francis Group, LLC

© 2024 G. Unnikrishnan and V. Pratapkumar

ISBN: 9781032328645 (hbk)
ISBN: 9781032328652 (pbk)
ISBN: 9781003317081 (ebk)

DOI: 10.1201/9781003317081

Typeset in Palatino
by codeMantra

For my grandchildren Sharan, Pooja, Kalyani, and Shankar who sometimes

show me that the world and projects are not as complex as we think they are.

G. Unnikrishnan

To my colleagues and professional associates from my trainee days up until now.

V. Pratapkumar

Contents

Part II   Case Studies and Lessons

Preface

Why a book on Front End Engineering Design (FEED) in oil and gas projects? The answers are several.

The progress of a project from concept to completion involves a large number of technical, financial, and management operations, performed by various specialist agencies. Broadly, there are 'deliverable' functions like design, engineering, estimation, procurement, construction, and commissioning, which are tied together by project management supported by functions like project controls, contract management, quality and Health, Safety, and Environmental assurance. Each of these operations and functions is critical for delivering a project successfully.

We (GUK and VPK) have together nearly four-plus decades of experience in engineering and project management of oil and gas facilities.

Over the years, we have seen that one of the major factors for time and cost overruns in oil and gas projects is the deficiencies associated with the engineering process, predominantly in the early stages of project conceptualization and definition, and prevalent to a lesser extent throughout the project lifecycle. These early-stage activities are the foundations of subsequent steps, and any gaps or errors at this stage will impact the downstream activities adversely. There was also a realization that the basic issues associated with such deficiencies can be corrected to a large extent by a change in approach by project owners, consultants, and contractors.

Oil and gas projects are no strangers to time and cost overrun. Historically, oil and gas projects have underperformed. However, such data were generally brushed under the carpet of high oil prices. In fact, a recent (2019) analysis of 500 completed projects above USD 1 billion, during the past 5 years, by a consultant, shows that 60% experienced schedule delays, while 38% had cost overruns.

Now, the volatile oil market, fluctuating oil prices, and alternate energy sources are posing serious challenges to the industry. For the industry to survive, there must be a significant control of the time and cost overruns and improvement in efficiency for new projects, with the associated benefits of minimizing duplication of engineering efforts, and eliminating protracted contractual disputes and litigation.

We had come across occasions where a little bit of care or supervision, some more reviews, and/or the right person for the job would have prevented major mistakes and goof-ups that erupted during Front End Engineering Design (FEED), detailed engineering, construction, or operation causing major financial loss. Assumptions made during FEED, instead of necessary engineering studies, often crash during detailed engineering.

A sound Statement of Requirements (SOR) from another entity in the owner company is often the basis for the FEED. Ambiguities, inconsistencies, and gaps in the SOR sometimes surface very late in projects and will affect the quality of FEED itself. A good FEED requires certain time, because it is a game of studies, verifications, complex multi-disciplinary reviews, and feedback to the design staff requiring spending manhours. The right amount of quality manhours spent on the job will result in better FEED.

Again and again, we have seen the owner companies' lack of quality in the data given to the consultant, who takes it in and rush to complete the FEED. Owner companies often specify unrealistic FEED time, to cover up the time lost elsewhere. Writing the requirements of FEED is a science, as well as an art. Unless written carefully, it will seriously affect the project's engineerability and constructability. There is no substitute for the quality of the data from owner companies and verification of the same by the consultant.

Detailed engineering is done usually by the Engineering, Procurement, and Construction (EPC) contractor and involves preparation of thousands of engineering documentation (called deliverables). These deliverables require approvals from the client and go through several rounds of reviews before it is finalized and stamped as 'Good for construction'. Once the detailed engineering phase is delayed or goes out of sync from the plan, it is very difficult to catch up.

Experts from several engineering disciplines develop the FEED and voluminous detailed engineering documentation. Individual disciplines generally do a good job of what is assigned to them. However, when they are taken together, sometimes their product develops flaws. It may contain errors, inconsistencies, mismatches, gaps, etc., just like the misspelt genetic code in the DNA sequence. Then, the project will have problems during detailed engineering, construction, and/or operation. In short, it becomes the case of individual excellence and collective failure!

There are several good books on project, engineering management, and design of oil and gas facilities. Nevertheless, literature on how to excel in FEED and engineering management with perspectives from both players of the game, that is, from the owners' and contractors' sides, is indeed unique. This book attempts to give the reader a sense of the conflicting interests and reasoning when a project is analyzed from the viewpoint of a project owner and a project contractor.

This book describes the more common pitfalls, errors, and blunders that the stakeholders make while delivering capital projects with examples and remedies. The viewpoints of the key players are sometimes diametrically opposite. However, the changing oil and gas market demands a more realistic, dynamic, and collaborative relationship, in order to survive. Unnecessary duplication of design efforts, contractual litigation, and expenses must be reduced. Otherwise, the industry may not survive at all! This book focuses on this aspect and presents such a unique view of the subject to the readers.

Though the book is based on the FEED and EPC model of project execution, the principles can be applied to other types of contract models.

This book is primarily meant for engineers and executives working in the oil and gas industry, related engineering consulting firms, and Engineering, Procurement, and Construction (EPC) contractors, especially in the areas of engineering, design, and/or project management. It will be useful to the following:

- The owner companies when they write the project Statement of Requirements (SOR) and prepare data for consulting companies (FEED contractors).
- The FEED contractors when they develop the project design basis, specifications and drawings, the deliverables, and the Tender Documents for the project contractors.
- The EPC contractors who bet their money on the job, to evaluate the FEED/Tender Documents for gaps and inadequacies, and to assess the risks arising from such inadequacies, when they prepare the EPC bid and later execute the project. It will also help them when formulating project execution strategies, plans, and budgets.
- The academia: Faculty and senior students in project, engineering and construction management, and allied disciplines can get practical insights into the development of FEED, which usually remain in the files, manuals, and procedures of companies. They will also get an insight into one of the major causes of project time and cost overrun, and its possible remedies.

G. Unnikrishnan and V. Pratapkumar

About the Authors

G. Unnikrishnan has over 40 years of experience in the oil and gas industry in India and Middle East. He retired as Engineering Specialist from a national oil and gas company. His experience spans the areas of process design, process safety, and engineering and project management. He advised companies on the quality of Front End Engineering Design and how understanding and measuring the same can improve project delivery.

He is currently working as a consultant to design and engineering companies in India and abroad. His expertise are mainly in process design and process safety and how process plant design and operations can be optimized to improve process safety. He is an active researcher in the area and is the author of a book on application of the Bayesian network methodology to risk assessment of oil and gas equipment. He has also presented and published papers on process safety in several international conferences and technical journals.

He is a certified Functional Safety Engineer on Safety Instrumented Systems. He holds a degree in Chemical Engineering from Calicut University, MTech from Cochin University of Science & Technology, and PhD from the University of Petroleum and Energy Studies, Dehradun, India.

V. Pratapkumar is a chemical engineer with over 40 years of experience in project implementation in the oil and gas, petrochemical, and fertilizer industries. His predominant areas of experience are in project management followed by engineering and construction. He started his career as a Management Trainee with Fertilizers and Chemicals Travancore (FACT) and has subsequently held senior-level positions in UHDE (India), P.T. Polysindo Indonesia, Pipeline Engineering (India), and Petrofac Engineering & Construction, UAE.

He retired from Petrofac in 2014 as Vice President and Head of Project Services. He is currently advising L&T Hydrocarbon Engineering as a consultant in project management competency development. His direct experience with project implementation is in India, Indonesia, Georgia, and Kuwait. He has also held oversight positions for projects in several other Middle East, Asian, and North African countries. His active involvement with EPC project implementation in the oil and gas industry since 2002 has given him valuable insights into the issues related to implementation of large projects, and the critical role the quality of Front End Engineering Design documents plays in the success of an EPC project.

Introduction

This is a book about Front End Engineering Design of oil and gas projects and how it can be done successfully ensuring an error-free and executable design. It is of utmost importance since oil and gas projects involve large capital investments, considerable efforts, and long period of time.

Project Management Institute (PMI) offers the definition of a project as follows:

'A project is **temporary** in that it has a defined beginning and end in time, and therefore defined scope and resources.

And a project is **unique** in that it is not a routine operation, but a specific set of operations designed to accomplish a singular goal. So, a project team often includes people who don't usually work together–sometimes from different organizations and across multiple geographies'.

By the above definition, we realize that the engineering team working on the project will also be on 'collaboration on need to basis' and will not stay together even during the design or execution of the project.

From the above fact, it is clear that the challenges in design and engineering of a project are primarily in the integration of the huge number of activities that are parallel, sequential, dependent, or interlinked. The engineering management of a mega project is no mean task. The engineering manager is confronted with several questions on a daily basis. The issues can crop up from many sides, from the upstream reservoir engineers who want to change the input data during midway of the Front End Engineering Design (FEED), from facility operations who want to relocate the facilities, to the civil design engineer who does not have sufficient data about soil to design foundations for the huge storage tanks.

In the race to meet the scheduled targets of completion, sometimes, assumptions are made, and decisions are taken, which in the later days during construction or operation may prove fatal to the project or constructed facilities.

The decision to launch the ill-fated Space Shuttle Challenger is a classic case of 'schedule pressure' overcoming engineering judgment. Some engineers in NASA knew that the polymeric O ring that sealed the solid booster rocket sections may lose its properties on that very cold launch day. But their voices were not heard by the management.

The design of any complex piece of technology, such as oil and gas processing facilities, refineries, aircraft, space shuttles, computers, nuclear plants, or submarines, all requires careful attention to the details, as well as the overall picture. Without the above, the chance of failure is quite high.

As mentioned in the preface, data tell us that failures are more common than success in upstream oil and gas projects, which require corrective action urgently. Given the post–COVID-19 situation and slow-down of the world economy, we cannot afford to be complacent and think of business as usual. There is an urgent requirement to deliver projects within budget and schedule, which are safe and reliable.

This book is an attempt in that direction. It does not elaborate on project and engineering management in detail. However, sufficient background on the above is given for completeness. It is based firmly on the authors' first-hand experience in management of mega projects in upstream oil and gas.

This book is structured in the same way as the project lifecycle, based on the stage gate project progress, from the identification of the business opportunity, Feasibility and Statement of Requirements (SOR), definition of the project through FEED, construction, and commissioning and handover. The terminology could be different in some companies, but the content and intent are the same for practical purposes. All the above sections relate to the perspectives and case studies from both owner's and contractor's sides, which makes interesting reading.

Certain overarching issues are also discussed. Among others, it will cover the following:

Understanding the project environment and Volatility, Uncertainty, Complexity, and Ambiguity (VUCA): Very few engineering and project managers understand the nature of the current VUCA that pervades throughout the project lifecycle. Personal bias, conceptual errors, changing multinational workforce, supply chain disruptions, contractual Gordian knots, etc. all pop up when least expected. Managing them and moving forward could become difficult.

Project risk assessment and management: It usually receives a confusing treatment. It seems everybody knows the risks to the project, but the majority of the actions taken end up being non-effective. Often, its use in projects is not up to the mark.

Various engineering studies, including Value engineering, Health, Safety, and Environmental (HSE) and Security studies: These are useful and productive tools, but the project manager is sometimes challenged with diluting the analysis or the results. When the project is a hundred-meter race, the above studies can slow it down.

Emotional intelligence – the art of managing the project team and stakeholders: The sensitive art of diplomacy is required to draw out the best performance of the engineering team. These men and women are specialists of extreme nature, and unless they carefully handled, the engineering deliverables can be affected in many ways. Emotional intelligence is the key to managing the players of the game.

This book is written in a concise style with minimum jargon and real-life examples keeping in mind the practicing project personnel. The case studies provided in a separate section will be of interest to researchers and students. We hope that this book will be of good use to the professionals and academia, in achieving excellence in design and engineering.

Part I

Project Lifecycle, Gate System, and Perspectives

1

Project Framework: Lifecycle and Gate System

A brief description of the project lifecycle is given here without going into details.

Readers will recall that a project has definite beginning and end in time. What is in between is the project lifecycle. The lifecycle is characterized by distinctive features from its beginning to end. The above features are conveniently classified into five stages. They are listed in Table 1.1, in the general terminology used in oil and gas, with an alternate common name in the brackets along with a brief description.

A detailed description of each of the above stages, covering their scope, implementation procedure, and potential challenges, is explained in the succeeding chapters.

Beyond the project definition stage described above, the majority of the large oil and gas projects are implemented using the EPC model. The reasons why this is a preferred model, and what the risks associated with such a model are, are explained in the next few paragraphs:

- EPC is an acronym for Engineering, Procurement, and Construction. It is a very simplified term, which covers a very wide range of activities and responsibilities related to project implementation. It defines a strategy of project implementation wherein a contractor undertakes the total responsibility for implementing a project starting from design to commissioning. The other common strategies for project implementation are Engineering, Procurement, and Construction Management (EPCM), Cost Reimbursable, Build, Operate and Transfer (BOOT), etc.

- In the EPC project model, a contractor takes full responsibility for completing a project for a fixed lump sum price. The risk of fully understanding owner's (client's-from the contractor point of view) requirements, and of accommodating all possible uncertainties inherent in implementing complex high-value projects is absorbed by a contractor, for a substantial reward. What the owner gets is the assurance of a project completion schedule, a project cost, and the quality of the completed project. For a contractor, it is a 'high-risk, high-reward' proposition.

TABLE 1.1

Project Stages

Stage Number and Name	Description
1. Identification	The original project ideas are explored and confirmed with respect to need, utility, possibility, and fit to the company's overall plans. Options are identified.
2. Feasibility and Statement of Requirements (SOR) (Assess and Select)	The basic technical, Health, Safety, and Environmental (HSE), and financial parameters of the proposed project options are studied to confirm its compliance with the various parameters (technical, HSE, financial, schedules, etc.). The most optimum option is selected, and a set of requirements (Statement of Requirements) is developed for the next stage of project.
3. Project definition (Define)	The project is 'Engineered' to give a clear definition of the scope of the project and its requirements. At this stage, the cost estimates of the project are refined and reconfirmed for financial viability, and the technical scope of the project is set as a benchmark for project construction. The resultant documentation from this exercise is the Front End Engineering Design (FEED). This activity can be carried out by the owner themselves inhouse, or by a specialist project management contractor engaged by the owner.
4. Construction (Execute)	FEED documentation and associated commercial and financial documents form the Invitation to Bid (ITB) package, which is tendered out. At this point in time the owner formally approves the expenditure for the project. Project cash flows are planned accordingly. The response to the ITB that is the bids, are received, evaluated, and clarified, and the contract award/s are finalized. On contract award, the primary responsibility would transfer from the owner to one or more project contractors. The contractor completes detailed engineering, the required studies, procurement of equipment and materials, site development, and all associated civil/structural construction works. The procured equipment and materials are installed and connected. Construction work related to all disciplines is completed.
5. Commissioning and handover (Operate)	The installed project is tested for safety and integrity, and the various components of the project are sequentially put into operation, in order to establish that the project can produce the desired output and quality as defined in the project definition stage and the contract.

- There are other possible implementation models, such as the Engineering, Procurement and Construction Management () model, the Reimbursable Cost model, the Cost Plus model, that have substantially lower involvement, responsibility, and risk, on a contractor's part, as compared to the EPC model.

- The single biggest differentiator between an EPC model and the other project implementation models is the degree of risk involved and the ability to manage those risks. The success of an EPC project is not based on merely 'taking the risks', but on a systematic process of identifying and managing the risks involved. Can every single potential risk be identified and managed? No, almost all can be identified, but not all of them can be managed. So, EPC contractors can at best try to mitigate the impact of the unmanageable risks.

In the earlier days, projects used to proceed from the stage to stage with general administrative, financial, and management approvals. Rigorous reviews between any of the stages with intention of differing or stopping the project were not very common.

A 'phased review process' was introduced and practiced in NASA in the 1960s for development of space missions. Later, with the advent of software projects the term 'stage gate process' became prevalent. 'Water fall process' is another term that is used to visualize work that cascaded down from a previous phase to a current phase as a series of mini waterfalls.

Essentially, it means that a project can proceed to the next stage only after a thorough review at the end of each phase. The reviews are usually conducted by a team from senior management outside the project with the required authority to take decision about the project. Relevant technical experts also take part in the review. In typical oil and gas capital projects, the decision involves any one of the following:

Proceed

    After review of the presentation by the project proponent and the submitted documentation, the stage gate review committee gives go ahead for the project to move to the next stage. This means that, at that point in time, the project proposal satisfies the overall plans and requirements of the company. In short, the business case for the project is strong.

Rework and resubmit

    If the project in general meets the requirements, but needs certain fine-tuning and minor changes, then the committee will recommend reworking the proposal. The committee will give definitive recommendations on the areas to be reworked and the desired outcomes. A timeframe for resubmission is also suggested for completing the rework and resubmission.

Reassess

    In this case, the committee has determined that the project does not meet the requirements and will request that the proposal need to be reassessed. Reassessment means that the concept itself has to be reviewed, which is different from the case of rework and resubmission, where the committee is satisfied with the project concept.

Stop

    The decision to stop the project is taken when the proposal does not meet the basic requirements and overall plans of the company at that point in time. It also means that the business case for the proposal is weak. It may be revived at a later date when the conditions become favorable.

The process can be visualized as depicted in Figure 1.1.

The same methodology can be used repeatedly for each of the stages resulting in the stage gate process for the entire project. Please see Figure 1.2. G1 to G5 represent the stage review gates.

As seen in Figure 1.1, the key element for the success of the stage gate process is a set of comprehensive and clearly written project governance procedures along with the structure and systems in place. Further, all project stakeholders have to be fully aware of the existence and compliance requirements of the above.

The key players in the project lifecycle are defined below:

The owner company: The owner company invests money on the project for definitive business and financial objectives. They own the project. On completion and operating the project, the owner company, ultimately makes profits or suffers losses due to the outcome of the project. In this book owner and the company has been used interchangeably for the term 'owner company'.

The project proponent: They are the department or entity in the company that proposes the project, based on the requirements and forecasts of their area. In oil and gas industry, it will be usually connected to the operations support, oil field development, and upstream planning sections. The project proponent is part of the owner company. The stages 1 to 3 are with the owner company who:

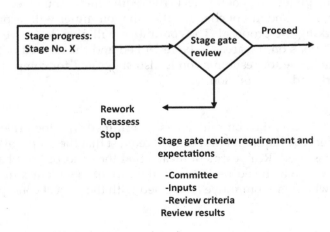

FIGURE 1.1
Typical stage gate progress and stage gate review.

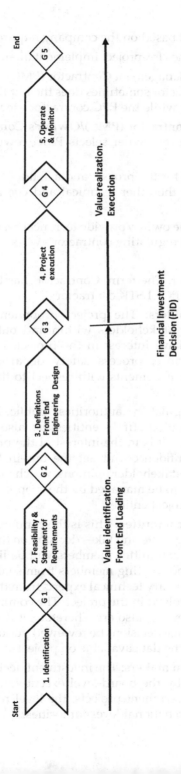

Figure 1.2

Project lifecycle and stage gates.

- Defines the project based on the company requirements.
- Arranges the finance for project implementation.
- Selects a Project Management Contractor (PMC) and an EPC contractor. PMC contractor sometimes does the FEED and supervises the overall project, while the EPC contractor executes the project.

Project Management Contractor (PMC)/Owner's Consultant

In the cases where the owner selects PMC/ owner's consultant, they

- Prepare the FEED for the project, and a Tender Document based on the FEED and the other technical and commercial requirements of the client.
- Act on behalf of the owner/provide technical assistance to owner in monitoring and regulating contractor's work.

The Contractor

- The EPC performer: The term 'Contractor' has been used interchangeably with EPC/ LSTK contractor

Other project stakeholders: The project proponent discusses the project with the key stakeholders within and outside the owner company, who will have interest in the project. For example, the operations, maintenance, process safety departments, etc. will want to have their requirements with respect to the lessons learnt included in the project.

Local/state/national regulatory authorities, public interest groups, external utilities, potential offtake entities, etc. also will have a certain say in the project. It is in the interest of the proponent to take stakeholders into confidence and get the buy-in from them. The internal dynamics of stakeholder opinions, which may have conflicts sometimes, will have to be managed by the proponent up to Stage 2 and then on by the project entities.

The stage gate review committee: This is the committee that reviews the project at each stage gate and takes decision on the fate of the proposal. Generally, the committee members will be limited to 7, with certain permanent and rotating members from senior management supported by the necessary technical experts from the company who are not otherwise involved in the project. The committee's decisions are not infallible. They can also err. Therefore, it is suggested that a review of the committee decisions be reviewed periodically, say once in 5 years or so with the data available on projects.

The investment decision makers: The investment decision on the project is usually made by the board-level executives. In certain situations, in case of government projects, the level required for such decision may go up to national-level authorities.

The project entities: They are the departments in the owner company who is responsible for executing the project and include the design, project management, construction management, and project control functions. The project entities will be led by the project manager, who will be the single point of accountability for the project. Their role usually starts from Gate G2, that is, once the project proposal is selected, and the Statement of Requirements (SOR) is finalized. However, sometimes, they are consulted very often during Stage 2 itself.

The project, design, and engineering units in the owner companies tend to be small these days, consequent to downsizing and retirements. Nonetheless, sometimes, they are entrusted with certain projects by the company (inhouse) to do the FEED. The project management basically supervises the PMC.

1.1 The Main Elements of the Project Gate System Structure

1.1.1 Inputs to the Stage Gate Review

The input documentation and studies needed along with the formats will be defined for each stage. They are the supporting elements for the proponent's case for moving the project forward. The project proponent presents the proposal to the review committee along with the supporting documents. The documents can be generalized to a certain extent for projects. However, the same needs to be customized for oil and gas projects. (For example, for building projects, there can be major differences). Even with oil and gas projects, there can be differences, which need to be highlighted.

1.1.2 Review Criteria

Availability of the required documents alone does not ensure soundness of the review. For an effective review, the right members, correct criteria, and sufficient time are essentially needed.

The stage gate reviews are always supported by the relevant technical experts in the company, who are not involved in the project. If such expertise is not available, it can be outsourced.

Criteria for stage gate review usually consist of general as well as specific. It is flexible based on the projects, rather than a fixed methodology for all types of projects. To make it objective, sometimes, there is a scoring system. More about this is given in the relevant chapters.

1.1.3 Review Results

Review results have been discussed earlier. The decisions of the review committee are communicated on record without any ambiguity so that the proponent knows exactly what to do.

1.2 Value Identification and Front End Loading (FEL)

The owner company spends time and money to identify a potential project during the initial stages of the project. All options and alternatives are considered, and the values delivered by the best options are evaluated during this stage. Sufficient and necessary details and documentation known as Front End Loading (FEL) must be available at this point. FEL has been recognized as one of the best industry practice and a key factor in ensuring success of a project. In summary, FEL requires that:

- Sufficient time and resources are allotted during the early stages of the project.
- The business opportunity is framed, fully explored, and evaluated.
- Risks are identified, and uncertainty range is minimized.

It may be difficult in some projects to judge the sufficiency of FEL. But with experience, the team and review committee can determine the situation. The review criteria itself may prove to be too generic.

The basis of all decisions must be trackable at the conclusion of gate G2.

The opportunities for adding value to the project diminish as the project progresses, as summarized in the typical S curves shown in Figure 1.3.

Readers can appreciate that, unless careful governance is exercised in all dimensions and aspects, projects can go terribly wrong.

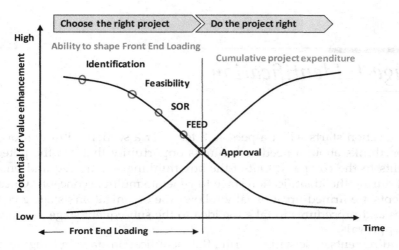

FIGURE 1.3
S curves – decreasing opportunity vs. increasing cost.

KEY POINTS TO NOTE

- The governance procedure for the Project Gate System should contain methods on how to handle special cases and exceptions. For example, a revamp project with a simple scope multiplied by n number of locations will qualify for stage gate process by its cost, but the scope itself may not merit the time and effort required for it to go through the stage gate process. However, exceptions should not become too common.

- Periodic review of the stage gate system with respect to the projects' performance will be beneficial to correct the deficiencies.

- Gate G3 project definition is the most important gate, when the owner makes the decision to commit to EPC contractor, often known as Final Investment Decision (FID). After gate G3, the opportunity for FEL is lost. Therefore, it is prudent to take time at this point to consult with project stakeholders once again (better still, conduct a workshop) and clear the pending issues, understand the risks, and finalize plans on how to manage the risks, if not done earlier. The decisions taken in the workshops should be recorded meticulously.

2

Stage 1: Identification

Identification starts when a person, a team, or a section within an organization thinks about a need or business opportunity that has the potential benefits to the company. Out-of-the-box thinking is required and encouraged during the identification stage to generate multiple concepts. Once the concepts are firmed up, it usually follows the organization's stage gate processes and procedures to take the idea to the subsequent stage gate review and approvals.

Requirements of activities during the identification stage in a highly summarized form are given below:

i. Frame the business opportunity (Rationale, objectives, and scope)

   A workshop with the stakeholders will provide the information or indicate the lack of the same and help the project proponent in developing business opportunity, background, scope, objectives, rationale, etc., clearly.

ii. Identify and engage key stakeholders and interfaces

   Identification and engaging all stakeholders at the identification stage itself are essential. Analysis of the stakeholder's inputs will provide information about possible interfaces with other projects and other activities within the company.

   Potential interfaces must be searched and identified in corporate targets, forecasts, infrastructure plans and schedules, etc.

   Such advance planning will bring out the aspects of coordination efforts that will be needed in later stages of the project.

iii. Conduct a preliminary technical evaluation

   The project proponent must carry out a preliminary technical analysis at the identification stage. It may appear very approximate, but it is needed to see the viability of concepts. It will also identify each capital project element, for example, land, facilities, utilities, infrastructure. Other parameters include the following:

   - Indicative figures on uncertainty of project parameters.
   - Applicable technologies.

DOI: 10.1201/9781003317081-4

- Health, Safety, and Environmental requirements, particularly local government regulations and laws.
- Operational requirements.

iv. For moving forward to the review for Stage 1, organizations usually will require certain formats for submitting the documentation. A summary of typical documentation requirements is given in Table 2.1.

Once the deliverables are ready, it is sent to the stage gate review committee for review and decision on the further course of the project.

The review committee usually consists of executives from other related departments/teams of the company. The committee structure will be

TABLE 2.1

Stage 1: Identification Stage – Typical Documentation

Key Deliverables		Content
Executive summary	Business opportunity proposal	• Business opportunity summary • Cost estimate (order of magnitude, −30% to +50%) • Benefit realization • Opportunity timeframe • Risks and issues • Requirements/plan for Stage 2 • Recommendations
Reference documents	1.0 Business opportunity statement	1.1 Introduction and summary 1.2 Description (objectives, scope, etc.) 1.3 Rationale (business drivers, objectives, SWOT analysis, strategic fit) 1.4 Benefits, risks, issues, and impacts 1.5 Key assumptions 1.6 Key success factors 1.7 Timeframe (required/expected) 1.8 Technical assessments 1.9 Additional information required
	2.0 Capital Investment Analysis	2.1 Cost estimate (order of magnitude, −30%+50% accuracy)
	3.0 Stakeholder management	3.1 Identification of stakeholders 3.2 Engagement requirements 3.3 Engagement action plan
	4.0 Risks and issues	4.1 Major risks (risk register) 4.2 Potential issues
	5.0 HSE compliance	5.1 HSE compliance and documentation requirements
	6.0 Lessons learnt	6.1 Lessons learnt (from previous projects)
	7.0 Requirements and Plan for Stage 2	7.1 Scope of work and schedule 7.2 Resource requirements 7.3 Contract funding requirements

flexible so as to have permanent members and members on a rotation basis. For example, members from project management, engineering safety, and finance teams could be permanent, whereas those from teams that have on part or intermittent involvement could be on a rotation basis. The review committee also has the mandate to call any expert from within the company for advice and recommendation.

The project proponent presents the project Stage gate 1 deliverables to the committee. The committee goes by written assessment criteria. Typical criteria used are given in Table 2.2. The criteria need periodic review to keep it relevant and up to date.

TABLE 2.2

Stage Gate 1.0: Identification – Assessment Criteria

Sl. No.	Stage Gate 1.0: Identification – Assessment Criteria	Assessment Review Comments
1.0	Review business opportunity statement	
1.1	Is the business need or opportunity purpose and objectives clearly understood?	
1.2	Did the assessment cover critical success factors and key challenges?	
1.3	Are the major outcomes Specific, Measurable, Realistic, and Timely (SMART)?	
1.4	Will introduction of the capital project have an impact on the timing and prioritization of other existing or proposed projects?	
1.5	Have the potential interfaces and other implications with other capital projects, ongoing projects and operations been identified/understood?	
1.6	Have the capital project development options been reviewed to adequately identify the scope for further studies?	
1.7	Will the proposed studies of the development options provide a sufficient definition of the requirements (e.g., production facilities, utilities, infrastructure) to confirm the technical viability of the business opportunity?	
1.8	Have the key project uncertainties been considered when assessing the capital project technical viability?	
1.9	Have the potential technology challenges been reviewed?	
1.10	Which technical aspects will require an in-depth investigation to be conducted during the Feasibility stage (Stage 2)?	
2.0	Review cost estimate	
2.1	Has an order of magnitude cost estimate of the required capital investment been provided?	
2.2	Has the preliminary capital cost estimate been compared against previous similar projects?	

(Continued)

TABLE 2.2 (*Continued*)

Stage Gate 1.0: Identification – Assessment Criteria

Sl. No.	Stage Gate 1.0: Identification – Assessment Criteria	Assessment Review Comments
3.0	Verify the identification and assessment of major risks	
3.1	Have major risks been identified, including preliminary mitigation approaches	
3.2	Is the risk exposure acceptable?	
4	Verify the identification of the key stakeholders	
4.1	Have principal stakeholders been identified, their roles and responsibilities confirmed?	
4.2	Have the required stakeholder engagement actions been initiated and if not what is planned?	
5.0	Verify HSE compliance requirements	
5.1	Have the HSE compliance requirements for this stage been identified and appropriate plans and procedures initiated and/or completed?	
6.0	Plan for the way forward	
6.1	Have the resources required for conducting Feasibility and SOR assessed?	
6.2	Will third-party personnel be required for Stage 2.0? If so, is a service contract in place for the same?	
6.3	How much contract funding will be required and what will be the respective phasing for these funds?	
6.4	Has the plan for Stage 2.0 been fully developed and documented?	

3

Stage 2: Feasibility and Statement of Requirement

Once Stage gate 1 review is over and the gate is closed, Stage 2 is triggered automatically. Here, the development options will be subjected to a structured evaluation process (Feasibility study) to assess technical, economic, risk, and HSE perspectives.

In some companies, for convenience, Stage 2 is further divided into two subsections, namely, 2.1 and 2.2. Stage 2.1 is Feasibility, and Stage 2.2 is Statement of Requirements (SOR). The same division is used in the summary given below.

3.1 Stage Gate 2.1: Feasibility

It will evaluate the options given from Stage 1 and narrow down the choice to one project. In doing so, it should achieve the following objectives:

1. Ensure all feasible concept development options have been reviewed and suitable conceptual designs prepared.
2. Ensure that relevant operations and maintenance requirements have been reviewed and incorporated into the conceptual designs and identified for development in future stages.
3. Prepare cost estimates and run incremental project economics to support and justify the selected concepts.
4. Ensure that all development risks have been identified and adequately reviewed.
5. Select the optimal concepts (projects) for further development during the next part of the stage gate process.

Probably the best way to achieve the above objectives is to conduct workshops/peer reviews with the identified stakeholders. The workshops/peer reviews should be planned properly so that all stakeholders can be present.

The project proponent prepares the documentation to satisfy the above and presents the proposal to the stage gate review committee.

DOI: 10.1201/9781003317081-5

A list of key activities required for completing Stage 2.1 is given in Table 3.1.

At this point of time, the project proponent shall forward a copy of the documentation to the project execution team (often called the project team) so that they can be familiar and ready to participate in the review of the upcoming Stage 2.2: Statement of Requirements (SOR).

TABLE 3.1

Stage 2.1 Feasibility- Key Activities

Sl. No.	Activity	Responsibility
1	Mobilize the Feasibility team (members from relevant stakeholders)	Project proponent
2	Develop a list of concept alternatives	Feasibility team
3	Evaluate and shortlist concept alternatives: The Feasibility team shall use a structured evaluation method for assessing each concept alternative. A set of suggested parameters are listed below: a. Technology b. Schedules c. Costs d. Risks e. HSE f. Economics Note: 3.1 The goal of the above is to apply an all-round and multi-disciplinary approach to the evaluation and comparison of alternatives under different and possibly conflicting perspectives. 3.2 Sometimes, it may not be possible to define a concept alternative that is superior to all its key parameters (e.g., technical feasibility, economic value, satisfactory risk, good business fit). The alternatives with the most suitable trade-offs shall be shortlisted as a result of the evaluation and comparison process.	Feasibility team
4	Perform sensitivity analysis Sensitivity analysis shall provide additional comparisons and test the remaining alternatives against different development scenarios. The scenarios shall consider the variation of such key parameters as Capex, Opex, production rates, and energy prices to allow modeling the expected impact on the program's business objectives.	Feasibility team
5	Identify and select preferred concept If required for making the final decision, the project proponent may use advanced decision-making tools (such as decision trees). The key outputs of the activities performed during this stage shall be included in the Feasibility report.	Feasibility team
6	Prepare the documentation as per the required format and submit to the stage gate review committee. A summary of deliverables is given in Table 2.1	Project proponent and Feasibility team

TABLE 3.2

Stage 2.1 Feasibility – Summary of List of Deliverables

Stage 2.1: Key Deliverables		Content
Executive summary	Business opportunity proposal	• Selected concepts (projects) • Cost estimates • Key economic indicators • Benefit realization – update • Risk and issues – update • Requirements and plan for the next substage • Recommendations
Reference documents	1.0 Business opportunity statement	1.1 Introduction and summary 1.2 Description (objectives, scope, etc.) 1.3 Rationale (business drivers, objectives, SWOT analysis, strategic fit) 1.4 Benefits, risks, issues, and impacts 1.5 Key assumptions 1.6 Key success factors 1.7 Timeframe (required/expected) 1.8 Technical assessments 1.9 Additional information required
	2.0 Capital Investment Analysis	2.1 Cost estimate (order of magnitude, −20%/+40% accuracy)
	3.0 Stakeholder management	3.1 Identification of stakeholders 3.2 Engagement requirements 3.3 Engagement action plan
	4.0 Risks and issues	4.1 Major risks (risk register) 4.2 Potential issues
	5.0 HSE compliance	5.1 HSE compliance and documentation requirements
	6.0 Lessons learnt	6.1 Lessons learnt (from previous projects)
	7.0 Requirements and plan for Stage 2	7.1 Scope of work and schedule 7.2 Resource requirements 7.3 Contract funding requirements

The key list of deliverables for the stage gate review for Gate 2.1 is given in Table 3.2, while a summary of the stage gate assessment criteria is given in Table 3.3.

3.2 Stage Gate 2.2: Statement of Requirements (SOR)

The latter half of the Feasibility stage is known as Statement of Requirements (SOR). In this stage, the selected concept/project is developed further as SOR, which includes the technical details, schematics, preliminary computer

TABLE 3.3

Stage 2.1 Feasibility – Assessment Criteria

Sl. No.	Stage 2.1: Feasibility – Assessment Criteria	Assessment Review
1	Review the quality of the Feasibility study	
1.1	Have the preliminary options identified in the previous stage been expanded and sufficiently detailed?	
1.2	Has an adequate range of alternative technical concepts been studied to take into account different options?	
1.3	Was an economic model utilized to compute profitability indicators and evaluate alternativeness of concept alternatives?	
1.4	Have the concepts to be carried forward in the final selection process been properly described?	
1.5	Have the most promising concepts been compared against the drivers consistent with business goals (e.g., economic value, timeline for project delivery, capital requirements, environmental impact, saturation of the spare capacity of existing facilities, piloting of new technology)?	
1.6	Has appropriate sensitivity analysis been performed to test the most promising concepts against different scenarios?	
1.7	Does the proposed concept optimize value and project trade-offs (e.g., recovery rate Vs reserves, Capex Vs Opex, cost Vs quality, risk Vs reward)?	
2.0	Assess potential impact of project risks that define preliminary risk control strategy and plan.	
2.1	Have all risks been identified and potential impact assessed?	
2.2	Has the risk control strategy been defined?	
2.3	Has the risk management plan been defined to assign risk ownership and formalize the monitoring mechanism?	
2.4	Have any critical project risks been identified that require immediate attention prior to proceeding to Stage 2.2?	
3.0	Confirm that key stakeholder expectations are understood	
3.1	Have key stakeholder expectations been reviewed and factored during the Feasibility stage?	
3.2	Have specific actions and accountabilities been planned to manage key stakeholder expectations?	
4.0	Verify Health, Safety, and Environmental (HSE) requirements.	
4.1	Have the HSE compliance requirements for this stage been completed?	
5.0	Plan for way forward	
5.1	Has the work plan for Stage 2.2: Statement of Requirements (SOR) been fully developed and documented?	

simulations and calculations, site selection and project area, preliminary environmental screening, updating of project risk assessment. The SOR as a minimum includes the following elements:

    i. Capital project scope and timeline.

    ii. Report on-site visits and site selected.

   iii. Plot reservation activities if any required with internal and external stakeholders.

   iv. Capital project data (feedstock, product quality, performance guarantee requirements, environmental emission limits, etc.).

    v. Design philosophy (modularity, expandability, sparing, etc.).

   vi. Conceptual designs (PFDs, plot plans, schematics, etc.).

  vii. Long lead equipment.

 viii. Material specification for major items.

   ix. Specifications for special equipment and procurement strategy.

    x. Requirements for any special services (market surveys, pre-qualification of Original Equipment Manufacturers (OEMs), specific engineering studies required during FEED, third-party inspection services, etc.).

   xi. Operability and maintainability considerations.

  xii. Shutdown and start-up requirements, both hardware and software, as required.

 xiii. Automation and control system requirements.

 xiv. Summary of Stage 1 and Stage 2.1 deliverables, as well as HSE deliverables.

  xv. Lessons learned from similar previous projects.

A formal capital project proposal is also prepared for information and approval of the line management. This is mainly because, after Stage 2, most companies will need external services for developing the project definitions/ FEED, whereas Stages 1 and 2 can be done generally inhouse, with external inputs on a need basis. Capital project proposals will contain requirements for carrying out the project definitions/FEED with respect to the project scope, resources, funds, phasing, schedule for completing the required project FEED, and timing for preparing a corresponding budget proposal. The project proposal includes the deliverables for Stage 2.2.

    The above two activities continue in parallel in most companies.

    To achieve the above objectives, the key items to be reviewed and confirmed during the reviews with stakeholders will be typically the following:

   a. **Capital program**. The proposed capital project includes all the required elements such as dependent projects in accordance with the company policies.

   b. **Stakeholder engagement**. All stakeholders internal and external to the company have been engaged and their requirements addressed in the SOR. It could include additional resources and funding for the stakeholders.

c. **Development options.** The development options such as interfaces, dependencies, potential overlaps, development timelines, etc. were identified and evaluated.

d. **Capital program risks.** The uncertainties and risks associated with the execution of the capital program have been identified and quantified in the risk management plan.

e. **Capital program economics.** The conceptual cost and the timelines for each capital program component (capital project) have been determined. The expected economic performance of the overall capital program has been evaluated in line with company policies and guidelines.

f. **Contracting strategy.** A preliminary contracting strategy for executing the capital program components (capital projects) has been developed in consultation with the project management groups of the company.

g. **Departments/teams.** At this stage of the project, the key players in the project execution have to be identified and the FEED preparation requirements discussed with them.

A peer review will be beneficial at the end of Stage 2.2. Such a peer review/ workshop can be conducted involving the concerned stakeholders, as well as other experts from the company. If external expertise is required, the same must be obtained.

Preparation of SOR is the responsibility of the department that proposes the project (project proponent). However, it would be best, during this stage, to involve the project execution (project management), project control, and contracts departments so that their inputs can be incorporated into the SOR. (As mentioned earlier, the project team will be already having a copy of the SOR by with time).

The project team has an opportunity at this stage to review the SOR and prepare their resource requirements, schedule, and budget estimates for capital project/s. If multiple projects are part of a capital program, each of the projects shall be taken up for the above task.

At this stage, a reasonable estimate, FEED schedule, and project schedule are prepared with the help of the above teams, and as noted earlier, a formal capital project proposal is prepared to go along with the Stage 2.2 deliverables for management approval.

A summary of Stage 2.2 deliverables is given in Table 3.4.

A summary of assessment criteria passing Stage gate 2.2 is given in Table 3.5.

The project team shall prepare the key documentation 'FEED execution plan' at this stage.

TABLE 3.4

Stage 2.2 Statement of Requirements (SOR) – Key Deliverables

Key Deliverables		Content
Executive summary	Statement of Requirement (SOR) proposal	• Project scope and development requirements • Cost estimates (−20% to +40%) • Key economic indicators • Benefit realization (estimates) • Schedules • Risk and issue update • Requirements and plan for Stage 3.1 • Recommendations
Reference documents	1.0 Business opportunity statement	1.1 Introduction and summary 1.2 Description (objectives, scope, etc.) 1.3 Rationale (business drivers, objectives, SWOT analysis, strategic fit) 1.4 Benefit, risks, issues, and impacts 1.5 Key assumptions 1.6 Key success factors 1.7 Timeframe (required/expected) 1.8 Technical assessments 1.9 Additional information required
	2.0 Capital Investment Analysis	2.1 Cost estimate (order of magnitude, − 20% to + 40% accuracy)
	3.0 Stakeholder management	3.1 Identification of stakeholders 3.2 Engagement requirements 3.3 Engagement action plan
	4.0 Risks and issues	4.1 Major risks (risk register) 4.2 Potential issues
	5.0 Contracting strategy	5.1 Review and recommendation of feasible contracting strategies[a] for the project.
	6.0 HSE compliance	6.1 HSE compliance and documentation requirements
	7.0 Lessons learnt	7.1 Lessons learnt (from previous projects)
	8.0 Requirements and plan for Stage 2	8.1 Scope of work and schedule 8.2 Resource requirements 8.3 Cost estimate for conducting FEED 8.4 Signed and approved CPP

[a] Review of possible contracting strategies is important at this stage of the project. To do the same, there must be written down procedures to evaluate the nature, characteristics, and parameters of the projects against the contracts that are best fit.

TABLE 3.5

Stage 2.2: Statement of Requirements (SOR) – Assessment Criteria

Sl. No.	Assessment Criteria	Assessment Review Comments
1.0	Review SOR and readiness for FEED	
1.01	Is the SOR complete in all respects to allow the preparation of an adequate FEED?	
1.02	Has the project proponent included the project management team/s in the review of SOR?	
1.03	Has the previous operations and maintenance experience been leveraged in developing the operations and maintenance requirements for the new project?	
1.04	Has the development of the preliminary contracting strategy involved all required stakeholders?[a]	
2.0	Update risk control strategy and plan	
2.01	Has the risk control strategy been updated?	
2.02	Have the project risks and potential impacts been identified?	
2.03	Have all the major risks been evaluated and specific mitigation actions and controls been identified?	
2.04	Have any critical project risks been identified that require immediate attention prior to proceeding to the FEED stage?	
3.0	Confirm that key stakeholder expectations are understood	
3.1	Has the key stakeholder expectation been identified, reviewed, and factored into the SOR development process?	
3.2	Have the specific actions and accountabilities been planned to manage key stakeholder expectations?	
4.0	Verify HSE compliance requirements	
4.1	Have the HSE compliance requirements for this stage been completed?	
5.0	Plan for way forward	
5.1	Has the FEED team structure been reviewed and the required personnel skill and availability been assessed?	
5.2	Will third-party personnel be required for execution of Process Stage 3.0. Is service contract and personnel available for the same?	
5.3	How much contract funding will be required, and how it will be phased?	
5.4	Has the plan for Process Stage 3.1: project definition/FEED been fully developed and documented?	

KEY POINTS TO NOTE

- Feasibility studies should apply the same set of criteria to alternative options to arrive at one single project to move forward.
- The Feasibility should evaluate both Capex and Opex.
- The impact of HSE issues should not be understated. Environmental regulations are getting tighter all over the world. Therefore, attention should be given for sustainable solutions to engineering problems. Green technologies should be evaluated.
- All technical/specialized studies required during the next stage shall be listed in SOR. The SOR shall also give directions as to what is to be done regarding the outcome of the studies in either case, whether it is positive or negative.
- SOR contents need to be a consensus decision by the project proponent and stakeholders from their associated departments. It should not be one person's ego trip. Once the SOR goes through the next gate to Stage 3, an associated stakeholder should not work against the project.
- Ideally, the project proponent along with its associate teams should present the SOR in a workshop. Important scope items, specifications, and the unknowns and risks should be discussed with the participants. Items to be resolved should be listed, and follow-up action and reporting should be meticulously done.
- Packaging the project, the right way is important. If the scope that is basically unfit for the EPC/ LSTK contract is packaged out as such, it is likely that the project will fail to meet its expectations and objectives
- External inputs and critical additional information required should be identified and obtained at this stage itself, rather than postponing to the FEED stage.
- A qualitative (still better, quantitative) evaluation of the quality of the SOR will be helpful for ensuring the completeness and soundness of the SOR. This subject is further discussed in a later chapter.

4

Stage 3: Project Definitions/Front End Engineering Design (FEED)

4.1 Objectives and Key Activities

The key objectives of Stage 3 are to:

- Provide a basis (known as Front End Engineering Design) for the preparation of the project execution plan and conducting the detailed engineering design.
- Refine the capital program/project cost estimates and schedules, in line with the level of accuracy required to raise program/project budget proposals.
- Confirm the soundness of the capital program's key success factors (strategic fit, execution capability, operations targets, economic viability).

Stage 3 responsibility is with project team/s of the company. They may be assisted wholly or partly by a Project Management Contractor (PMC). At the end of Stage 3, the definition of the scope, cost, and schedule and execution strategy for the capital projects (and the overall capital program) must achieve the level of definition required to make a sound investment decision in line with the company's strategic operational goals and financial requirements. Stage 3 is subdivided into Stage 3.1: definition; and Stage 3.2: project endorsement and approvals.

The completion of FEED stage activities and respective deliverables shall demonstrate that:

- A clear definition of the objectives for the capital program/projects is available.
- Execution plan is valid and aligned with the company's strategic objectives.
- Cost and time estimates are defined with the required level of accuracy.

DOI: 10.1201/9781003317081-6

- Key stakeholders have been engaged in Front End Loading and endorse the execution.
- HSE and quality management requirements are incorporated into the project execution plan.

A summary of key activities during FEED is as follows:

1. Mobilize the FEED team (inhouse or outsourced).
2. Detailed review of Statement of Requirements (SOR).
3. Conduct FEED and finalize contracting strategy.
4. Development of project scope, and all deliverables agreed with the stakeholders.
5. Finalization of cost (to the required accuracy) and schedule (to the required level).
6. Update the economic model.
7. Update and finalization of preliminary project execution plan.
8. Finalization of all required documentation for Stage 3 review by the committee.

4.2 Contract Models

There have been cases where major projects have been developed with FEED to suit the EPC/LSTK contract, only to find later that that the model is not feasible for all the component units of the projects. This is an important factor where owner companies and project proponents must pay greater attention. There are various contract models available other than the LSTK/EPC model. For instance, in BOOT type of contracts, the contractor takes very high risks technically and financially and the chances of poor project execution and performance are comparatively higher. Therefore, proper due diligence studies and risk assessment must be done by both the owner (before finalizing the contract type) and the contractor, before submitting the bid.

The contractor will be responsible for all technical details, evaluations, and inputs, as required for the owner to confirm financial commitment. Common contract models are summarized in Table 4.1 along with its basic characteristics.

Apart from the above, owner companies now have several choices for outsourcing engineering and management services to expert companies/consultants. It has pros and cons for the owner companies. The main advantage is the small number of engineering personnel and resources that need to be on company's rolls. The disadvantage is that the capability for design, engineering, and management of projects will be no longer with the owner

TABLE 4.1

Contract Models for Project Execution

Sl. No.	Contract Model	Main Phases and Responsibility: O – owner; C – contractor						
		FEED	Det Engg	Procurement	Construction	Commissioning	Op & Maint.	Final Ownership
1	Lump Sum Turn Key (LSTK)	O[a]	C	C[b]	C	C	O	O
2	Engineering Procurement & Construction (EPC)	O	C	C	C	C	O	O
3	Engineering Procurement & Construction Management (EPCM)	O	C	C/O[d]	C/O[d]	C/O[c]	O	O
4	Cost plus	O	C	C/O[e]	C/O[e]	C/O	O	O
5	Open Book Estimate convertible to EPC	O	C	C/O[f]	C/O[f]	C/O	O	O
6	Re-Measurable	O	C	C	C	C/O	O	O
7	Build, Own, Operate, and Transfer (BOOT)	O Functional specs	C	C	C	C	C	O[g]

[a] In LSTK, the owner may sometimes go for a lesser definition of FEED, leaving the contractor to take the responsibility for fuller FEED definitions, while specifying the guarantee requirements.

[b] Sometimes, to meet the long delivery schedule of critical equipment, design and detailed engineering of such equipment are done during the FEED stage itself. The datasheets, specs, and purchase requisitions are finalized, and the order is placed formally on the selected bidder after a bidding process. The order is then transferred to the LSTK/EPC contractor through a 'Novation' agreement. Novation is a process by which contractual rights and obligations are transferred from the owner to the EPC contractor, with the consent of all parties concerned, and the existing contract is replaced with the new contract. It is basically a triparty agreement between the owner, the supplier (OEM), and the EPC contractor. It is called the 'Deed of Novation'

[c] In EPC, the owner may sometimes do commissioning by themselves or outsource it to a specialist agency. The contractor will be required to provide commissioning assistance by way of equipment and manpower.

[d] In EPCM contracts, the contractor takes the responsibility for detailed engineering of the project and procurement of major equipment if so required, and also performs the overall project management and construction supervision of the project. The EPCM contractor does not do the construction or take full responsibility for delivering the project on time and within budget.

[e] Cost plus is a contract where all the hardware costs are reimbursed to the contractor by the owner. The contractor is paid a service fee for doing the work related to the above. It can be fixed or variable based on the manhour cost. The construction and installation costs are paid separately to the contractor.

(Continued)

TABLE 4.1 (*Continued*)

Contract Models for Project Execution

f   Open Book Cost Estimate convertible to EPC (OBCE): In the OBCE method, the owner gets the FEED done either inhouse or through contractors. The second contract will be on an OBCE basis where the full detailed engineering cost estimate is done by the contractor. The contractor is paid for the above. Up to this phase, it is similar to the EPCM contract. The installed cost of each item including the breakup of contractor's profit is disclosed to the owner. On the OBCE basis, the owner's and contractor's transactions are fully transparent. The owner reviews the cost, technology, and OEMs of equipment and has a chance to modify/amend the equipment/specifications and selection and influence the cost. Once the detailed engineering and the cost are final, the owner company has the choice of converting the packages of their choice as EPC contract with the same contractor. Otherwise, the project will continue on the OBCE basis.

The method of incentivizing the contractor is important in the OBCE method. Instead of a manhour basis or a fixed percentage of the project cost, where the contractor may tend to inflate the cost, the contract terms often will be a fixed amount plus a percentage based on the contractor's success in optimizing and minimizing the project cost. Therefore, it will be in the interest of the contractor to work for optimizing the project and minimizing the cost during detailed engineering.

If conversion to EPC is envisaged, a certain time must be allowed in the schedule to cater for the same.

g   There are several variations possible within BOOT, some of them are as follows:

- Build and then operate on a separate service contract for a limited time. There are no specific guarantee requirements from the contractor during the operation and maintenance service contract period except the availability of personnel and competent operation and maintenance of the plant. The contractor does not own the facility.

- BOO, on a purely rental basis, in which the contractor has to dismantle and take away the plant after the limited time. The plant is owned and maintained by the contractor. They have to meet the output and quality guarantee requirements based on the inflow fluid streams and utilities provided by the owner. FEED and detailed engineering are by the contractor based on the functional specifications provided by the owner.

- BOOT, where the owner has the option to buy the plant at a previously agreed price after a limited period of operation by the contractor. FEED and detailed engineering are by the contractor based on the functional specifications provided by the owner.

- There are a number of project funding methods available now, including the formation of Special Purpose Vehicle (SPV) for the project, which eases the capital project expenses for the owner.

- Except for the methodology using SPV, mentioned above, all other BOOT types may not be suitable for large capital projects.

companies. Important competencies such as design, and engineering and project management capabilities will be lost from the owner companies. It is a decision that must be carefully taken.

4.3 Project Complexity

In making choices on undertaking the project execution, one of the key factors to be considered is the project complexity. There are several references available on the subject. However, whichever criteria are chosen to measure project complexity, they should be practicable.

Scoring system of project complexity should involve at least the following criteria as a minimum:

1 Lifecycle cost of the project.
2 Lifecycle duration of the project.
3 Project organization size and hierarchy.
4 Technology readiness.
5 Risk during project execution and operational phase.
6 Project financing options.
7 Visibility of the project owner.
8 Project owner's authorization basis for the project.

Based on the scoring obtained for each of the above parameter, the projects can be classified as high, intermediate, or low in terms of complexity.

4.4 Contracting Strategy

Following are the key points to keep in mind while thinking about contracting strategies:

i. Contracting strategy for the project should be thought of sufficiently early, preferably during Stage 1 identification itself. The review of potential contract models must involve key stakeholders, and contracting and legal experts. Such reviews must be documented and progressed through Stage 2 Feasibility and finalized by the end of Stage 2.2 SOR. FEED should be aligned to the contracting strategy and model selected.

ii. In order to help in choosing a strategy for project execution, a matrix of both project cost and complexity can be considered. Please see Table 4.2.

iii. The matrix will consist of project complexity on vertical axis of the matrix and project cost on the horizontal axis.

iv. Project complexity classification criteria is summarized in section 4.3. Further, project can also be categorized as small, medium, or major based on the capital cost and schedule of the project.

v. Once the scoring of the both the project cost/schedule and project complexity are finalized, go to the matrix given in Table 4.2 and select the horizontal axis scoring, for example suppose the scoring classifies the project in 'Medium'. Next see the vertical axis for project complexity. Suppose the complexity scoring places the same project in 'Intermediate'. Then the intersection of the vertical and horizontal axis give the possible strategy as 'Outsourced FEED. Execution by contractor-EPC basis'.

TABLE 4.2

Contracting Strategy Options Available Based on Project Cost/ Schedule and Complexity

		Project Cost/Schedule Score		
		Small	Medium	Major
Project Complexity score	High	Outsourced FEED. Execution by contractor-EPC basis.	Outsourced FEED. Execution by contractor-EPC basis	Basic FEED LSTK execution
	Intermediate	Inhouse or outsourced FEED, execution by contractor-EPC basis	Outsourced FEED. Execution by contractor-EPC basis	Outsourced FEED. Execution by contractor-EPC basis
	Low	Inhouse FEED. Execution by contractor (construction only or EPC basis)	Inhouse or outsourced FEED. Execution by contractor-EPC basis	Outsourced FEED. Execution by contractor-EPC basis

Note: 1. Smart outsourcing is the key to managing the entire project lifecycle.
2. Intersection of score of Project cost/schedule with that of Project complexity will highlight the possible contracting strategy.

vi. Whereas, if both the project cost/schedule and complexity are 'small' and 'low' respectively then a suitable strategy could be 'Inhouse FEED. Execution by contractor (construction only or EPC basis)'.

vii. The actual numbers are not mentioned in the book for the above scoring systems, since companies have different values, based on the individual company policies.

viii. With the above information, the owner can review the options available for implementing the project. Table 4.2 summarizes the options available.

ix. Do not throw unrelated scope items into a huge EPC project. Such unrelated scope will prove to be hurdles toward smooth execution of the project. For example, when the scope of a major pipeline project is being defined, it will be unwise to add to the project, scope on certain unrelated work inside process facilities in another location.

x. It is always better to divide a huge mega project (usually greater than USD 1 billion) consisting of different types of scope into suitably framed and scoped subprojects, with well-defined interfaces. For example, a mega project involving the construction of several storage tanks, pumping systems, aboveground pipelines, submarine pipelines, and marine structures, if put together as a single-bid package, could look easier initially from owner's view, but prove to be difficult later when it comes to getting competent EPC contractors who specialize in all the above. Depending on the scope, cost, and availability of specialist contractors, such scope can be divided into several smaller EPC contracts each dealing with one particular type of scope, for example, pipeline design, engineering and construction in one package, marine pipeline and structures in another package.

The main issue highlighted in such an approach is the effort for coordination required for handling and interfacing several projects simultaneously. The owner can always appoint a Project Management Contractor to oversee, supervise, and manage the coordination and interfacing effort. In such cases, adequate interface engineering must be done and included in the scope before sending out the Invitation to Bid (ITB). The money spent on such an arrangement will be worthwhile.

xi. Some owner companies keep accumulating changes and modifications inside their facilities, and when it becomes too complicated in scope and high in cost, they put it out as an EPC package. Firstly, such changes and modifications should have been undertaken as part of routine Management of Change and implemented as such, as and when the need arises. Secondly, putting it together as an EPC package without adequate field engineering to define the data, scope, and content has potential for a high number of disputes with the contractor later on.

4.5 Stage 3.1: FEED Deliverables

A summary of the documentation required for completion of FEED (Stage 3.1) is given in Table 4.3.

TABLE 4.3

Stage 3: Project Definitions/Front End Engineering Design (FEED) – List of Deliverables

Stage 3 Key Deliverables		Content
Executive summary	FEED proposal	• Project scope • Project execution strategy • Cost estimates (-10% to +15%) • Key economic indicators • Benefit realization • Overall project master schedule • Risks and issues • Recommendations
Reference documents	Project execution plan	1.1 Introduction (project description and background) 1.2 Overview of the project including objectives, scope of work, battery limits, cost, and schedule. 1.3 Execution plan (organization, governance, project controls and procedures, contracting strategy, construction methodology, etc.) 1.4 Procurement plan and strategy for long lead items and bulk materials 1.5 Overall project schedule 1.6 Cost estimates 1.7 Project monitoring and control 1.8 Quality management 1.9 Risk management plans and procedures 1.10 HSE compliance 1.11 Communication management

(Continued)

TABLE 4.3 (*Continued*)

Stage 3: Project Definitions/Front End Engineering Design (FEED) – List of Deliverables

Stage 3 Key Deliverables	Content
	1.12 Performance management
	1.13 Supporting plans (testing, pre-commissioning, commissioning, operations and maintenance, training and staff development, project close-out, etc.)
2.0 Capital Investment Analysis	2.1 Cost estimates (−10% to +15% accuracy typical)
	2.2 Economic evaluation
	2.3 Draft budget proposal
3.0 FEED package	3.1 FEED deliverables for all disciplines
	3.2 Utility requirements and deliverables
	3.3 Specific, Measurable, Achievable, Realistic, and Timely (SMART) project execution goals.
4.0 Site implementation plan	4.1 Approvals and permits
	4.2 Site selection and reservation
	4.3 Site risk analysis
	4.4 Plot plan
5.0 Operations and maintenance strategy	5.1 Operations requirements and strategy
	5.2 Maintenance requirements and strategy
6.0 Contracting strategy	6.1 Procurement plan for long lead equipment
	6.2 Procurement plan
	6.3 Contracting plan including selection of a particular contracting model and potential bidder list.
	6.4 Change management procedure
7.0 Stakeholder management	7.1 Stakeholder management plan
	7.2 FEED close-out meeting and minutes of meeting and sign off.
8.0 Risk and issue management	8.1 Risk register updated
	8.2 Issues and resolution
9.0 HSE compliance	9.1 HSE deliverables pertaining to Stage 3
10.0 Lessons learned	10.1 Review of lessons from previous projects
	10.2 Recording of lessons learnt in current project stage activities
11.0 Plan and requirements for Stage 4	11.1 Scope and schedule
	11.2 Resource requirements

4.6 Summary of Assessment Criteria

The assessment criteria for Stage 3.1 are given in Table 4.4.

TABLE 4.4

Stage 3: Project Definitions/Front End Engineering Design (FEED) – Summary of Assessment Criteria

Sl. No.	Stage 3.0 Assessment Criteria	Assessment Review
1	Confirm readiness for the Final Investment Decision (FID)	
1.1	Is the project scope still aligned with the business strategy?	
1.2	Has the project design been sufficiently developed to minimize the project execution risk?	
1.3	Do the project design and execution plan provide sufficient benefit realization to support the execution of the program/project?	
1.4	Does the updated economic evaluation continue to support an investment decision?	
2.0	Review of the FEED stage	
2.1	Have the required engineering disciplines been involved in the development of the FEED package?	
2.2	Is the project design detailed enough to achieve a cost estimate with the required accuracy?	
3.0	Review the contracting strategy	
3.1	Have the required stakeholders been involved in the preparation of the contracting strategy?	
3.2	Is the project execution plan detailed enough to identify the most appropriate contracting scheme for the work packages?	
4.0	Confirm the project scope, cost, and schedule	
4.1	Have scope, cost, and schedule been sufficiently defined enough for preparing a draft of the contract requisition?	
4.2	Are the project scope, cost, and schedule detailed enough for preparing a draft of the contract requisition?	
5.0	Review and update the project execution plan (PEP)	
5.1	Has the PEP been prepared?	
5.2	Does the PEP accurately outline the execution strategy, the project organization, staffing, and the contracting strategy?	
5.3	Have all the necessary procedures and control systems to support management and coordination of the project execution activities been developed and documented?	
6.0	Update risk management plan	
6.1	Has the risk management plan been prepared?	
6.2	Have all the major risks been evaluated and specific mitigation actions and controls been identified?	
6.3	Have any critical project risks been identified that require immediate attention prior to proceeding to Stage 4?	
7.0	Update stakeholder management plan	
7.1	Has the stakeholder management plan been updated, and have key stakeholder expectations been reviewed and addressed during the FEED stage?	

(Continued)

TABLE 4.4 (*Continued*)

Stage 3: Project Definitions/Front End Engineering Design (FEED) – Summary of Assessment Criteria

Sl. No.	Stage 3.0 Assessment Criteria	Assessment Review
7.2	Have specific actions and accountabilities been updated to manage key stakeholder expectations?	
8.0	Verify HSE compliance requirements	
8.1	Have the HSE compliance requirements for this stage been completed?	
9.0	Plan for the way forward	
9.1	Do the PEP, cost estimates, and schedule provide a solid basis for the preparation of a budget proposal?	
9.2	Has the plan for the next Stage 4, project execution/construction, been fully developed and documented?	

4.7 Stage 3.2: Project Endorsement and Approvals

The main objectives of Stage 3.2 are to:

- Confirm that the project planning processes such as project execution plan, HSE plans, resource plans, risk management plan, stakeholder management plan have been endorsed and will properly support the required project execution activities.

- Ensure that the cost estimates as to the required accuracy (usually −10/+15% accuracy) are required. The cost estimation is completed and placed on record usually called 'Tender Estimate'.

- Ensure that all the other project execution business activities have been completed in line with the required initiation of project execution (e.g., regulatory approvals, mobilization of resources, internal or as Project Management Contractors (PMC)).

- Ensure that the required commercial, financial plans are in place to support the initiation of the contract action activities.

- Conduct due diligence on the ITB package and prepare a due diligence report. The due diligence report is best prepared by another team (or third party) that did not participate in FEED or preparation of ITB.

- Initiate action for the required financial and management approvals. Usually, the approvals follow the sequence given below:
  - Initial approval for the budget proposal.
  - Final Investment Decision (FID) or Approval for Financial Expenditure (AFE).

- Initiation of contract requisition.
- Approval of bidder list.
- Final Invitation to Bid (ITB) package (with review and inputs from the contracts team).
- Confirmation of financial approval.
- Issue of Invitation to Bid.

The project management and associated design and engineering teams are usually given a final chance to review and verify the bid package before it is issued out.

The next steps in the process involve the following:

- Receive bids.
- Conduct technical and commercial bid evaluation. (This is usually done by a bid review committee, specially constituted for the task from the relevant depts of the company along with the contracts team. There will be pre-approved bid evaluation criteria and plan.)
- Bid clarifications and confirmations with the bidders.
- Bid evaluation report and Recommendation for Award.

Note: Some companies go for a two-bid system, wherein the bids are received in two separate parts (envelopes), one part containing the technical and commercial aspects and other containing only the price.

The price bids are opened only for those bids that have been found suitable after evaluating technical and commercial bids. Price bids for unsuitable techno-commercial bids are not opened at all, or in some cases, revised price bids are invited from all parties based on a uniform techno-commercial basis indicated by the project owner after review of all the technical and commercial bids. The advantage of the above method is that it provides a fair and transparent bid evaluation model for all participating bidders.

4.8 Key Points to Note in Stage 3

1. FEED is a key stage, after completion of which the company takes the step forward to commit money and resources to go ahead with the project.
2. Engineering management has an important role to play during this stage. Leadership and resources need to be available at this stage.

The engineering manager position must be manned by a suitable person with adequate qualification and experience.

3. Schedules need to be finalized to provide adequate time for conducting various engineering studies and safety studies. It must be understood that these studies cannot be done in a rush. Some of the studies have major impacts on the FEED and scope of the project itself. Expertise and resources must be planned for such studies.

4. Engineering activities should be planned carefully not to end up with HSE studies being on the critical path!

5. Resolving issues with stakeholders is a critical activity. The best way to resolve is to have a numbered, coded and listed resolution sheet with status updated periodically for each stakeholder and issue, during the progress of FEED engineering work. Any unresolved issue should be settled through workshops with participation of experts both internal and external. No issue should be left unresolved, because it could be raised up again by an unfriendly stakeholder during detailed engineering of EPC contract execution. Such issues could also become a bone of contention between the project owner and the EPC contractor during the implementation stage, resulting in project delays/cost overruns.

6. It is important that FEED should have the right balance. The FEED strategy should be to specify the right parameters only and leave the optimization part to the EPC contractor and Original Equipment Manufacturers (OEMs). Ideally, FEED should not be over specified or underspecified.

7. Engineering studies undertaken during the FEED should have clear objectives and recommendations based on which the project management and stakeholders can take decisions. It should not be something like a research study, which should have been done during the Stage 2 SOR.

8. Methods of improving the productivity of engineering (value improvement practices) must be implemented and pursued continuously. Engineering activities should not find itself in a situation where the reviews, resolution, and incorporation of stakeholders comments take longer than the engineering itself!

9. Scope creep or ballooning is a common feature in almost all projects worldwide. It will be worth going into details of the same for all owner companies. The project management must be vigilant for scope creep and must set up a Management of Change (MOC) system during FEED for tracking any scope changes and resolving them.

10. Cost estimates should match with engineering scope inclusive of the latest revision. It is common to see costing made on a previous revision of engineering documentation.

11. Any critical engineering study, which can significantly impact the scope or specifications of a project as defined in the FEED, should not be left for the project implementation stage, as this could adversely impact the project cost and schedule.

12. The bid package is a set of documents that is to be understood by a third party, who was not involved in preparing the FEED and who would be unaware of any unspecified assumptions or criteria. It should basically be clear and comprehensive when read by such a third party. The above issue of readability, clarity, and comprehensibility of the bid package is the single most important factor that determines competitive bidding.

13. For major and mega projects, it is definitely worth spending time and money in conducting a biddability or due diligence review on the ITB package, preferably by a third party.

4.9 View from the Other Side: Perspectives from the Bidder

1. From the point of view of the EPC bidder, the FEED represents a comprehensive and accurate detailing of all the critical requirements of the project owner. What is not clearly indicated in the FEED is expected to be developed/optimized by the EPC bidder within the limits of applicable International Engineering Standards/Good Engineering Practices.

2. The general observation is that the FEED put out in most EPC Tender Documents tends to be either over specified or underspecified. Both these situations are potentially detrimental to the smooth EPC implementation. The over specified FEED specifies a lot of detail developed within a short timeframe and with a minimum of resources. Such details are likely to change during the detailed engineering phase of the EPC implementation, and such changes can impact the cost and schedule assumptions made by the EPC bidder at the bid stage. The result would be protracted negotiations between the project owner and the EPC contractor leading to delays and cost overruns. Also, in many cases, an overspecified FEED restricts an EPC contractor from developing technically superior and/or more cost-effective alternatives, which could have benefitted all stakeholders.

3. An underspecified FEED could leave a lot of gaps in the owners' essential requirements and a lot of ambiguous specifications, which can have different interpretations at the time of detailed engineering. EPC contractors tend to plug such gaps and ambiguities with 'fit-for-purpose' solutions, as against a project owners' expectation

of a 'best-in-class' solution. The end result would be disputes and protracted negotiations leading to delays and cost overruns.

4. It is recognized that it might not be possible in some cases for an owner to develop a comprehensive and accurate FEED, due to limitations in resources, time, or in some cases a lack of information/data. In such cases, the contracting model must be changed to accommodate the possible changes in the future, due to detailed engineering developments, in areas where the FEED is incomplete. The alternate contracting models in such cases can be a 'Quantity re-measurable with fixed unit rates' model or an 'Actual cost plus service fee' model. An interesting case study of a project, which used a combination of EPC and a 'Quantity re-measurable with fixed unit rates' model, is given under case studies.

5

Stage 4: Construction

5.1 Stage Description

The construction or project execution stage as it is sometimes called of a project is the stage where the responsibility of completing the project passes on from the project owner to a contractor, or a group of contractors. The stage involves execution of a set of inter-related and integrated activities, and deliverables within a required timeframe. Typically, it consists of the following:

- Mobilization of the contractor.
- Mobilization of owner's project execution team.
- Detailed engineering.
- Procurement, inspection, Factory Acceptance Testing, and transportation.
- Offsite fabrication and transportation.
- Site preparation and on-site construction and installation.
- Mechanical completion and testing.
- Pre-commissioning.
- Commissioning.
- Start-up operations leading to Stable Operation.

The owner has already defined his technical requirements in the FEED document prepared by him, stipulated his commercial requirements, and put these together in a Tender Document, which forms the backbone of the contract for project construction. The general structure of a Tender Document for project construction would be as follows:

- A preamble, which states the broad objectives of the project, the details of the project owner, and the nature of the final contract that is envisaged.
- The commercial conditions of the contract, which specify the commercial rules and regulations applicable to the contract,

 DOI: 10.1201/9781003317081-7

governing laws applicable to the contract, contract price format, terms of payment, guarantees and liabilities, dispute resolution methodology, contract termination conditions, force majeure definitions, etc.

- The technical scope of work in the form of drawings, specifications, and descriptions.
- The Health, Safety, Security, and Environmental protection requirements that are applicable to the project.

In most cases, the owner would require that all contractors who are bidding for project construction comply with the Tender Document conditions fully, without any deviations. Such a condition would ensure a level playing field for all competing contractors, and also provide the basis for an easy and fast decision-making by the owner. The alternative approach of allowing competing contractors to assume and specify their preferred deviations will result in a complex and time-consuming techno-commercial analysis by the project owner to decide on the best option. This process can also come under criticism for subjectivity, favoritism, etc.

The following key activities will be conducted during the start of Stage 4: construction:

i. Mobilization of owner's and contractor's project teams.
ii. Nomination of focal points (person responsible for replying to communications and submittals from the contractor) from owner's internal departments.
iii. Holding the project kickoff meeting.
iv. Review and confirmation of internal and external procedures for managing the project interfaces.
v. Review and confirmation of project baseline schedule.

Other activities include the following:

1. Project execution plan, covering aspects such as detailed project implementation schedule, project management organization structure and mobilization plan, format and procedure for document flow between owner, contractor, and any relevant third party, change management procedure, progress measuring and invoicing procedure.
2. Detailed engineering, where the technical scope of work specified in the contract is elaborated to provide the specifications and documents required to purchase all equipment and materials required for the project, for installing and connecting all such equipment and

material, and for providing project infrastructure such as roads, drains, area lighting, compound walls.

3. Purchase and delivery of all equipment and materials required for the project. This process starts with the preparation of specifications and commercial terms and conditions for purchase, and goes through obtaining bids from approved suppliers, evaluation of bids and selection of supplier, monitoring and stagewise inspection of supplier's manufacturing work, Factory Acceptance Testing (FAT), transportation of finished goods to project site/offsite fabrication yards, storage, and issue of equipment and materials at the project site for installation. Approval of certain vendors' detailed engineering documentation also will be part of the above.

4. Site infrastructure development works such as clearing and leveling, compound walls, roads and drains, area lighting, site offices and stores.

5. Installation and interconnection of all equipment and materials at site, including construction of foundations, support structures, access ladders and platforms. Post-installation cleaning and testing of installed components for certification of integrity and safety.

6. Pre-commissioning and commissioning of the total installed facility in a phased manner in accordance with pre-determined sequence. At the end of this operation, the facility is deemed ready to commence commercial production.

7. Submission of operations and maintenance manuals for all the equipment and systems in the facility.

8. Training of owner's personnel as specified in the contract.

9. All through the processes described above, the contractor, along with his assigned subcontractors, is the performing agency, and the project owner and/or his assigned agents monitor, review, and certify the contractor's work in accordance with a pre-determined plan.

5.2 Deliverables and Approval Cycles

For a major industrial oil and gas project, the number of deliverables produced by a contractor at the construction stage runs to several thousands. About 80% of these deliverables would be engineering deliverables produced at the detailed engineering stage, and the balance 20% would be purchase-related documentation, and project management plans, procedures, and reports. At the start of the construction stage, a

Document Control Index is prepared, which will list all the deliverable documents produced by the contractor, with issue/approval dates for the various iterations of each document (issued for approval, issued for purchase, issued for construction, issued for information, etc.). This is the key document based on which the flow of information on a project is planned and monitored.

The procedure for approval of project deliverables by the owner and the approval cycles have a significant bearing on the schedule of a project. Out of the several thousand documents that a contractor generates, the number of documents that the owner wants to review can vary from a small number to a very large number. The number of documents for approval will basically depend on two factors, viz., the aspects of the project that the owner wants to closely control (as an example, the control and safety features of the process units, the layout of the facilities, the specifications for critical equipment, etc.), and the resources available with the project owner for reviewing and approving technical documentation. Having a larger number of documents for approval does not in any way change the respective responsibilities of the owner and contractor. What it does is to create a correspondingly large interface, which in turn presents increased risks of delays.

Once the list of documents for approval is agreed between the owner, the cycle of approvals will be, to a large extent, dependent on the quality of the FEED documents. A FEED that contains a lot of ambiguities, or one that has inadequate coverage of owner's requirements, will result in a misalignment between owner and contractor, and will consequently lead to multiple cycles of issue of related documentation, and in extreme cases, it can result in a dispute between the parties with respect to the scope of work. In either case, there would be significant delays to the progress of work on a project, and in extreme cases, there could be extended delays and potential cost impacts. Another reason for excessive cycling of documents requiring approval is the large number of agencies involved in the approval process from the project owner's side. Many of these agencies would not be fully familiar with the project contract conditions, and their requirements and preferences might not be consistent with the scope of work defined in the contract. A solution to this problem is that the owner should have all concerned stakeholders in the project to be familiar with, and in agreement with, the contract conditions before the contract is finalized. The above topic is highlighted in the previous chapters.

The approval of contractor's documentation by project owner or his nominated agent does not dilute the responsibility of the contractor, nor does it increase the liability of the project owner. The value it adds is that it acts as a 'check and balance' of the contractor's engineering efforts, and as a safeguard for the critical interests of the owner. If the approval process is oriented toward the above goal, then it will certainly add great value to the project.

5.3 Activities Toward Completion of the Project

Similar to the number of activities required at the beginning of a project, a sequence of critical activities is required to be executed toward the end of the project. These activities require advance planning, proper stakeholder management, and resource mobilization including vendors' engineers and commissioning experts by the contractor. A summary of typical activities is given in Figure 5.1, together with a brief explanation.

Usually, owner companies have definitions for the terms associated with mechanical completion, pre-commissioning, commissioning, and punch lists to be completed at each stage. Further, all of the above activities may not be taken up in certain projects. The following gives a summary of the above terms in general based on which the owner company can define their requirements, if they do not have definitions. It is important that definitions are part of the contractual terms.

5.3.1 Mechanical Completion

Mechanical completion will be achieved when the facility systems and structures, or portions of the works, have been installed and tested for its system fitness, in accordance with the final approved drawings and specifications and applicable codes, regulations, and laws, and are ready for starting the pre-commissioning activities.

Before the mechanical completion is declared, the owner's team goes through the facility and checks all aspects. The resulting observations are recorded as Punch List A, which needs to be completed, before mechanical completion can be declared.

Once the mechanical completion is declared, the Mechanical Completion Certificate is signed by both parties; the facility is ready for pre-commissioning.

5.3.2 Pre-commissioning and Construction Completion Certificate

Pre-commissioning activities will deliver a functional facility without the introduction of hydrocarbons, permanent energizing, or pressurization of systems. Pre-commissioning activities will be conducted after the approval of the pre-commissioning procedure submitted by the contractor to the owner after all category 'A' punch list items have been completed.

At this point, all 'as-built' drawings, operations and maintenance manuals, training program, etc., should be delivered to the owner. Most owner companies insist on delivery of the above, before the certificate of the next stage is issued.

A second punch list called Punch List B will be prepared to capture any gaps and issues noticed during pre-commissioning.

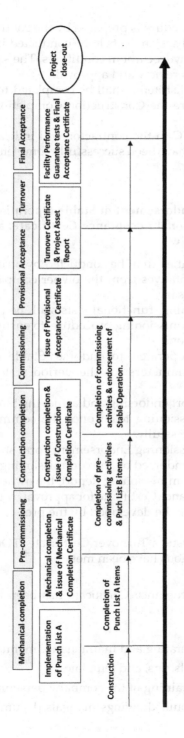

FIGURE 5.1

Activities toward project completion.

The commissioning procedure is prepared by the contractor and describes in detail the activities and system tests to be conducted in the process units, under real or simulated (hydrocarbon) conditions. The same is submitted to the owner company for its review and approval.

All category 'B' punch list items shall be completed to the satisfaction of the owner company before the Construction Completion Certificate (CCC) will be issued.

The issuance of the CCC to the contractor will indicate that all the pre-commissioning activities have been successfully completed and the commissioning activities can start.

5.3.3 Commissioning, Endorsement of Stable Operations, Issuance of Provisional Acceptance Certificate, and Turnover Certificate

The commissioning activities shall be conducted by a commissioning team that will include representatives from the owner company, contractor, and vendors, as deemed necessary.

During the commissioning, functional tests shall be performed in accordance with a detailed commissioning procedure, to ensure the facility meets the operational specifications.

The facility shall be operated to endorse 'Stable Operation', within the design operating parameters for the period stated in the Contract Documents.

A Commissioning Report endorsing and achieving Stable Operation shall be issued by the commissioning team once the commissioning activities have been completed successfully.

An appropriate Commissioning Endorsement Form shall be developed by the owner company and endorsed by the commissioning team.

Upon approval of the Commissioning Report, the owner's project team will issue a Provisional Acceptance Certificate for approval by the operations team.

An appropriate form will be developed by the project team and endorsed by the operations team.

The project team will issue a Turnover Certificate (TOC) to the contractor after the following conditions have been met:

- The Provisional Acceptance Certificate has been approved by the requesting team.
- The contractor has
  - Delivered all operating and maintenance manuals.
  - Delivered all tools, jigs, and templates.
  - Completed the training of the company personnel.
  - Delivered all 'as-built' drawings, manuals, documents, and records.

The issuance of TOC shall indicate that all requirements have been satisfied in accordance with the Contract Documents and the facility is ready for performance testing.

5.3.4 Capitalization of the New Asset

The project team will issue a Project Asset Report (PAR) to the project proponent for review and approval.

The project proponent is usually required to review and respond/approve the PAR within a reasonable time so that the asset can be capitalized.

The owner company's Financial Group requires a final, approved PAR to finalize the cost capitalization for the new asset in line with the owner company's procedures.

5.4 Key Points to Note

Potential Pitfalls and Bottlenecks

There are hundreds of reasons why projects are not successfully completed, ranging from incompetent contractors, project owners running out of money, etc., to natural calamities disrupting project implementation, government policy changes rendering projects unviable, etc.

For the purpose of this book, our analysis is based on large projects promoted by financially sound companies and managed by competent contractors, with no apparent adverse factors that can derail the project. And yet, such projects can fail, or can end up with huge cost and time overruns.

A recent example was the following report in a leading business paper of India, *The Economic Times* of January 4, 2019:

ONGC'S OFFSHORE DRILLING CONVERSION PROJECT DELAYED BY 70 MONTHS, COST OVERRUN OF RS 715 CRORE (USD 71.5 MILLION)

A project of ONGC, the country's (India) largest producer of oil and gas, that involves converting its oldest offshore drilling rig – Sagar Samrat – into a Mobile Offshore Production Unit (MOPU) has been delayed by more than 70 months and has faced cost overruns to the tune of Rs 715 crore – a status report by Ministry of Statistics and Programme Implementation (MOSPI) showed. (Authors note: original contract value of approx. Rs 860 Crores).

The delay in deployment of the MOPU has been one of the main reasons for the company's shortfall in crude oil production for financial year 2016–17 and 2017–18, oil ministry's petroleum and natural gas statistics report stated.

EY (Ernst & Young): Oil & Gas megaproject overruns to cost industry more than $500 billion

A new report by EY finds that 64% of oil & gas mega projects continue to exceed budgets, with 73% missing schedule deadlines. On average, current project estimated completion costs were 59% above the initial estimate. In absolute terms, the cumulative cost of the projects reviewed for the report has increased to $1.7 trillion from an original estimate of $1.2 trillion, representing an increase of $500 billion.

This was a project of national importance by the state-owned company, which was the largest producer of oil and gas in India and contracted to a competent agency. The exact reasons for this particular failure are not available in the public domain, but a case study of a similar project elsewhere is given later in this chapter, which will reveal some chronic issues in such large projects. This is not an isolated instance, in one specific country.

This is a worldwide phenomenon, as demonstrated in the study by Ernst & Young in August 2014, summary of which is given in the second box above.

In the authors' view, the primary reasons for misalignment between the owner and the contractor are as follows:

a. Inadequate or inaccurate information provided by the owner, based on which the contractor prices and plans his work.

b. The conflict between the inadequately articulated expectations of the owner and the time and cost limitations imposed on the contractor.

c. Inadequate understanding of the contract terms and conditions by the contractor.

d. Inadequate preparation by the contractor when planning for work in unfamiliar projects and geographies, remote locations, unfamiliar local labor laws, and so on. In one situation, the processes and procedures for getting gate passes to owner's sensitive sites took so long that it upsets the project schedule, a factor the contractor had not factored in at all in his planning!

e. It is the bidders' responsibility to do due diligence on the final bid submitted. If the bidder finds the owner's FEED and bid documents full of inconsistencies and gaps, it is better to declare it in writing to the owner rather than taking a gamble to move forward, which benefits nobody.

f. Protracted bid review and clarifications: If the bid review and clarifications take a long time, it is a sign of something is wrong with the ITB and the bids. When the owner extends the due date for bidding several times due to various reasons, it affects the back-to-back commitments made by bidders to equipment vendors, as well as the construction rates and assumptions made by the bidder. Sometimes, the time factor affects the owner's thinking itself. There are cases where during the intervening time, the project proponent revised the scope and technical requirements to such an extent that senior management ended up canceling the project, for sheer difficulties in getting finance approvals due to major scope changes. The owner company personnel should remember that major/mega project scope and technical requirements are the end product of teamwork and collaborative effort over a long period time. Unless the exchange of information and collaboration within the owner's stakeholders are sustained during the project execution, the project will get affected. Infighting within owner's stakeholders is often a cause for project delays.

5.5 Perspectives – Owner Mindset and Viewpoints

1. **There is no fool-proof ITB**: There is often thinking among many owner companies that their FEED and bid documents are ironclad and perfect. In fact, it is far from the truth. No FEED and ITB can be perfect. The FEED is the product of the choices exercised by the owner's project team along with stakeholders. The contract terms and conditions are, in 90% of the cases, one-sided and written in favor of the owner. Errors and gaps happen; unforeseen external events and time factors come into play. Severe weather, pandemics, and government policy changes are examples. There must be room for negotiations and resolution of disputes in a fair manner, and both parties must realize it. The authors' suggestion is that both owner and contractor include such mechanisms in the contract conditions itself.
2. **Concession Requests**: Due to several factors, during the execution of the project, the contractor is often compelled to raise 'Concession Requests' (CRs) to the owner. While CRs arise due to technical reasons, its impact could end up with cost and/or time implications. They are required due to conflicts among standards (between year of publication, international standards and company standards, obsolescence of the equipment specs vis-à-vis availability in the market (that too sometimes region- or country-wise, etc.)).

The owner must respond to CRs as soon as possible. If a response to CRs has to be practically possible, it is better to set up an inhouse multi-disciplinary team, with members from stakeholder departments to handle such cases. CRs should not go through the routine organizational review process since any delay in response may have an impact on the project schedule itself.

The owner company should also track all the CRs from all contractors. In the author's (GUK) experience, a simple database of all CRs will help in the above matter. By querying the CR database, the owner company personnel can find out which clause /specification/s and/or requirement is the root cause of the CRs and can take actions to rectify /fix them so that it is not repeated. The results will be surprising sometimes.

There were cases where a single clause in the owner company standard had given rise to a large number of CRs over many years in several projects, and nobody knew about it. This is because, in large owner companies, there will be several project teams working independently and stakeholder personnel involved may not be the same. They will be unaware of similar CRs in another project that popped up at different times. Querying such CR database should be a mandatory requirement during the due diligence process before the ITB is issued out. The CR database should be set up, administered, and updated by a department outside the controlling project teams. All concerned owner company personnel can be given access to query the database, generate reports, and forward to the relevant department/s for action.

3. **Micromanaging by owner companies**: The owner should not try to micromanage the contractors' tasks and personnel. Day-to-day construction issues at the site shall be managed by the contractor. While the owner company needs to monitor the contractors' personnel and site for security reasons, that should not end up in interfering with contractors' responsibilities.

4. **Lack of resources during detailed engineering and construction**: When basic design issues are raised by the contractor during detailed engineering and construction, it should be responded to promptly by the owner. The owner's problem will be lack of resources since most often, the FEED contractor who did would have left. It will be a good practice for the owner to have an agreement with the FEED contractor, to have their concerned personnel available during the

detailed engineering and construction phases of the project. A dedicated team consisting of FEED contractors' personnel and owner company stakeholders will be the best. The authors are aware of the logistics and practical problems of acquiring manpower resources, particularly in cases where the contract is awarded after a long time from FEED completion. In such cases, there must be careful advance planning to bring onboard a competent engineering contractor. Sufficient time also must be given to the engineering contractor's personnel to familiarize themselves with the project and the ITB documentation.

Similarly, when there are claims and adjustment orders to the contract, it gets resolved and settled and forgotten. It will be beneficial to the owner company to develop a database for all Adjustment Orders (AOs) to the original contract populated with relevant details. Again, querying the AO database will reveal the errors, gaps, and inconsistencies in contracts and will give the company a chance to study and rectify them.

5. **Project histories:** Projects are completed and forgotten. Owner companies are notorious for their short memories of projects. There is no dearth of technologies that can capture, store, and enable efficient retrieval of the thousands of documents generated during a typical project. But still, in most owner companies, the situation is bad. There were cases where an important HAZOP report could not be retrieved by the owner just 1 year into operation of the facility. It was required to verify critical hazard scenarios when certain modifications were proposed by stakeholders.

   It will be worthwhile to invest in a proper document storage and retrieval system by the owner companies. Such a system should be able to store not only the 'as-built' documentation, but also all the documents finalized for the project. [Paper storage is out of question!]

   Investing time and effort in studying completed projects will be ultimately beneficial. The usage of advanced analytics in all aspects of projects will reveal what went wrong actually.

   The topic is covered in detail under Chapter 9.

6. **Project close-out report:** The project team's job is not over, until the project close-out report is completed and presented to senior management. This is a forgotten task in most owner companies. Even if it is done, it will not contain anything worthwhile for a reader who is going to read it after 4 or 5 years and trying to glean important information from the report.

The authors' suggestion is to create systems and procedures within the owner companies to enable the project team to deliver a good quality report efficiently. For example, a good format for the project close-out report can help in the production of the report. Further, software-based systems and procedures can be in place for inputs to the report. Project close-out report delivery must also be part of the project quality management system.

Naturally, lessons learnt must be an integral part of the close-out report. It is the project manager's prerogative to carefully review, weigh, and include the correct topics that are lessons learnt. There is general confusion in owner companies regarding as to what is to be included as lessons learnt. It will be definitely worthwhile to spend some effort in developing guidelines for what is to be included as lessons learnt in the report.

7. **Project and engineering management training:** In owner companies, at least some of the personnel are accidental project/engineering managers. They are thrust into the role when they are least expecting or prepared for it. It shouldn't be so. Statistical studies have shown that project and engineering managers must have a good grounding in their own discipline before moving into the management roles. Oil and gas projects are increasingly becoming complex, and the owner companies should pay close attention to whom they are entrusting the mega project delivery.

6

Quality of FEED: Measuring It and Managing It

FEED quality can be measured. There are research projects that have proven that, when you measure the quality of FEED using the right parameters, it can predict with reasonable accuracy the outcome of the project, that is, its probability of success or failure. The objective of this chapter is to show that systems for measuring Front End Loading and FEED exist and are constantly improving. Due to the proprietary nature of the systems, only summaries are given here. Interested readers can look up the references given and see the consultants' websites for information.

Independent Project Analysis (IPA) is a consultant who measures the Front End Loading (FEL) in which FEED is a significant part. IPA defines FEL as 'the process by which a company develops definition of a project that was initiated to enable the company to meet its business objectives'.

Another agency, Construction Industry Institute (CII) views FEED as part of Front End Planning (FEP) and defines FEP as 'the process of developing sufficient strategic information with which owners can address risk and make decisions to commit resources in order to maximize the potential for a successful project'. Further, they define FEED as 'a component of the Front End Planning (FEP) process performed during scope definition (Stage 3), consisting of the engineering documents, outputs and deliverables for the chosen scope of work'.

The FEED package typically also includes cost estimates, schedules for the project, project execution plan comprising overall plans for detailed engineering, procurement, construction, installation, and commissioning.

As more and more data are collected on various projects, properly classified, and subjected to measurement systems along with statistical analysis and validation, the predictability becomes better. The following sections will summarize the tools already available and suggest certain methodologies that owner companies can use to start measuring and calibrating the FEL and FEED quality for their own projects.

6.1 Available Tools

The tools available currently are proprietary tools, developed by consulting companies. Some offer their services, while others offer the software and

TABLE 6.1

Quality of FEED: Measurement Methodologies

Consultant	Scoring Methodology	Scoring Range
Independent Project Analysis (IPA)	Front End Loading Index	3 (best) to 12 (worst)
Construction Industry Institute (CII)	Maturity and Accuracy Total Rating System (MATRS)	1 (worst) to 100 (best)
European Construction Institute (ECI)	Tasks and best practices for each task	–

train the owner's personnel to use the same. The quality of FEED measurement methodologies along with the respective consultants and the key features of their products are summarized in Table 6.1.

As can be seen, all the methodology's core concept is a scoring system for various parameters in FEL. The measured parameters can go up to several hundred. The measurement system also includes certain factors that measure the project team dynamics and organizational efficiency while handling major projects. Furthermore, the scoring system is different for various industries, which makes it more relevant. The other aspect is the validation/calibration of the scoring system with real-world projects. All the above consultants have calibrated their scoring system to various extents with data from real projects using advanced statistical tools. The methodology requires tracking of the project progress to evaluate and provide feedback on the measuring system, and the consultants and owner companies agree on a protocol for reporting the same. The greater the number of projects on which the technique is applied and tracked, the more accurate the predictability becomes.

A summary of each methodology is given below.

Independent Project Analysis (IPA): IPA basically reviews the development and deliverables from the initial three stages of the project, that is, Front End Loading 1 (FEL1), which is equivalent to the identification stage, stage 1; FEL2 equivalent to concept selection and Feasibility, stage 2; and FEL3 being equivalent to project definition/FEED, stage 3.

The key factors, varying from three to five, in each of the above stages are broken down into subfactors and further into various parameters, which are assigned points on a scoring system. The scoring system has weightages of each parameter, subfactor, and factors, and the weightages are adjusted based on IPA's assessment of the importance of each of them on the overall index. The scoring system and weightages remain proprietary. IPA's FEL factors, which are similar to the qualitative assessment criteria under each stage given in this book, are given in Figures 6.1a, b, and c.

Additional factors IPA considers are the alignment of functions (team practices), leading technology, project controls, and use of value improvement practices (VIPs).

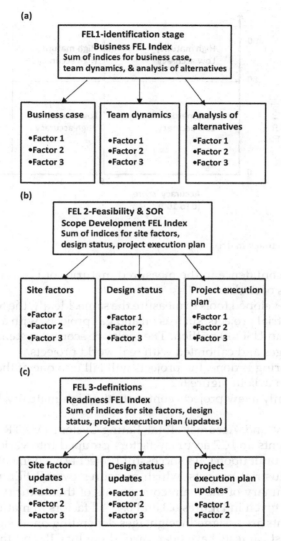

FIGURE 6.1
(a) IPA methodology – FEL1 Index, business case. (b) IPA methodology – FEL2 Index, scope development and (c) IPA methodology – FEL3 Index, readiness.

The overall scoring range is from 3 (best) to 12 (worst), and IPA notes that a score between 3.75 and 4.75 is the best practical range that the project should try to achieve.

Construction Industry Institute (CII): CII's techniques focus on FEED and define two key parameters, namely, maturity and accuracy to measure the quality of FEED. Maturity score measures 'the degree of completeness of the deliverables to serve as a basis for detailed design'. Accuraccy score evaluates

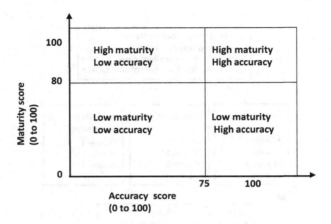

FIGURE 6.2
CII's maturity – accuracy matrix.

'the degree of confidence in the measured maturity of FEED deliverables to serve as a basis of decisons.'

They have developed tools to measure the same. Ideally, the tools are suited for large industrial projects, such as oil and gas projects, with a total installed cost greater than USD 10 million. The tools are scoring systems with appropriate weightages and calibrated with real-world projects.

Once the scoring is done, the projects will fall into one of the categories as shown in the matrix in Figure 6.2.

Obviously, only a few projects come into the high-maturity, high-accuracy category.

The Maturity and Accuracy Total Rating System (MATRS) contains 46 maturity elements and 27 accuracy factors grouped into various categories. The MATRS is built upon CII's successful Project Development Rating Index (PDRI) for industrial projects, which measures overall FEL versus project success. A summary of the main components of the maturity and accuracy assessments is given in Figures 6.3 and 6.4. All factors in maturity and accuracy components are assigned weightages depending on its significance.

Various statistical tests have been carried out by CII, and the results indicate that projects that score higher on the CII MATRS, maturity, and accuracy score have a higher rate of success.

European Construction Institute (ECI): ECI, as early as 1995, had published a book titled 'Total Productivity Management Vol I Guidelines for the Conceptual phase' to help personnel to check if they have complied with certain key factors. The important tasks ECI envisaged in the conceptual phases were as follows:

- Consents and permits.
- Project definitions.

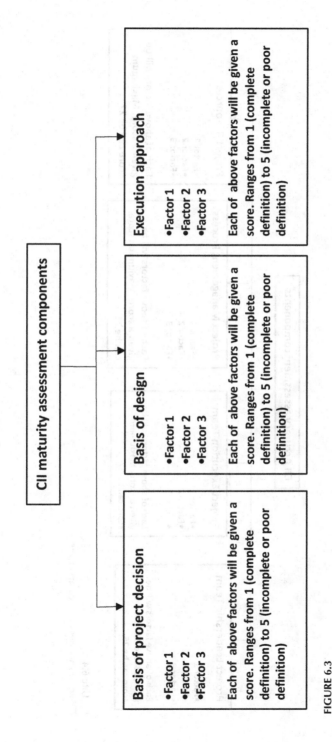

FIGURE 6.3
CII maturity assessment components.

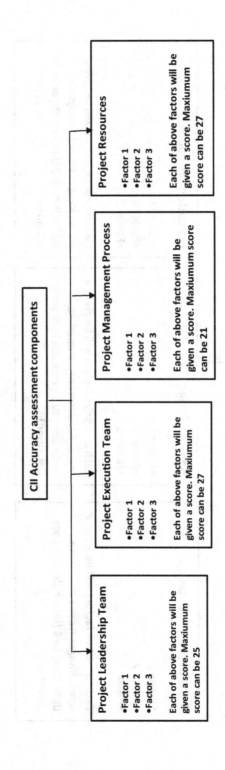

FIGURE 6.4
CII accuracy assessment components.

- Financial strategy.
- Project strategy.
- Contract strategy.
- Project management organization.
- Construction philosophy.
- Procurement strategy.
- Design of permanent works.
- Design of temporary works.

Each of the above tasks was then taken up in a Benchmarking format and split into several factors called 'Best practices'. These in turn have a scoring system 1 to 4 for 'Best practice were adopted' to 'Best practices could not be adopted'.

The scoring system results give an indication where the gaps are and what needs to be improved.

As can be seen, basically all the above measurement systems on quality of FEED recognize that there are certain key elements during the FEL and in the FEED, which cannot be overlooked and should be executed to the highest quality. If they are not done, there is a good probability that the project will suffer.

The above finding should be an eye-opener for the owner companies. A major project is an endeavor involving a combination of business, technology, and engineering. All of them must succeed to call a project, success. FEL and FEED costs come to about 2% to 5% of the project cost in most oil and gas projects. Unless sufficient time is allowed for FEED, there are chances that errors and mismatches can occur. Errors, mismatches, and gray areas will cost dearly to the owner and the contractor in the later stages of a major project. These have been proven by research.

Project success is defined typically as meeting the requirements of project outcomes, namely, project cost and schedule, and production capacities and qualities along with construction safety. The authors would add the safe operational performance of the facilities for an initial period of 3 years, also, as a measure of project success. Any one of the above cannot be traded off against the other. It is an iron frame.

If an owner company has not yet embarked upon measuring their FEL/FEED quality by proprietary methods, the authors' advice is to do so after the owner themselves has done a preliminary assessment of their FEL/FEED and improved it. Proprietary methods require time and effort and are generally rigorous. The proprietary consultant's initial assessment of FEL/FEED tends to be disheartening to most owners and their consultants.

In a simplistic view, the process of FEED and ITB production can be viewed as a set of systems that processes the inputs and produces the outputs as shown in Figure 6.5. It must be remembered that there are several iterative reviews and revisions of the documents in the process.

To enable an initial assessment, the authors have included under Chapter 5 a list of deliverables to be finalized in Stage 3. There are also assessment criteria provided to assess whether the FEED can proceed further as part of the stage gate review process. What is important is that the above verifications and assessments should be carried out objectively by a third party within the owner company who is not associated with the project and who has management experience and technical knowledge and competence. The third-party leader should have the acceptability of the stakeholders for her integrity.

6.2 Additional Factors

Additional activities that the authors would advise will be carrying out workshops to review the Frint End Loading (FEL) and its sufficiency. The suggested workshop and its timings are given in Table 6.2.

6.2.1 Standardization

Standardize or perish will be a motto worth remembering and applying to oil and gas projects. Many oil and gas projects suffer higher costs and delays due to the customization of equipment for the project on insistence from owner companies. Even equipment for the same duties differs from project to project within the same company! It is the responsibility of the engineering manager to actively promote standardization across the organization and projects.

Analysis of the system requirements, what is available in the market, and detailed inputs from OEMs will be required to go through the process. When the project involves procurement of a large number of one equipment, for example, compressor, the old saying 'Do not put all eggs in one basket' applies. It will be prudent to confine the selection to at least two suitable types so that their performance can be measured over a long period of time.

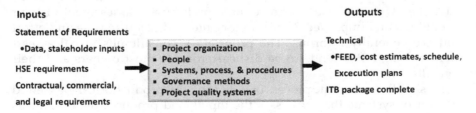

FIGURE 6.5
Production of FEED & ITB as a set of systems.

It saves time, effort, and money to repeat a successful design that is currently operational in a facility, with minimum changes. Please remember the Space Shuttle of NASA, which was basically designed only once and operated several times. It even had a reusable hardware part!

The initiative of the International Association of Oil and Gas Producers (IOGP) ongoing from 2016 known as Joint Industry Programme 33 (JIP33) is a welcome step in this regard. They have already published 55 procurement specifications with collaboration from the oil and gas industry, consultants, and OEMs. JIP33 specifications are based on the existing industry standards and provide full requirements to purchase equipment and packages. The published standards are free and can be accessed at the JIP33 website (https://www.iogp-jip33.org).

Other main factors that affect the quality of FEED are given in Chapter 7.

TABLE 6.2

Workshops[a] for Reviewing Front End Loading

No.	Timing	Workshop Subject	Participants	Objectives
1	After finalizing the SOR, before submitting for stage gate review	Defense of the project proposal by the project proponent with their team.	- Project proponent - Company project management and design unit - FEED contractor - Stakeholders - Other specialists	To understand: - The justification for business case of the project. - Basic technical scheme and requirements - Data sufficiency, inputs, and plot requirements - Main interfaces - Output and guarantee requirements Project execution plans - Risks identified at this stage
2	At 90% completion of FEED[a]	FEED completion overall review by the company project management team and design unit	- Company project management and design unit - FEED contractor - Project proponent - Stakeholders - Other specialists	To review and check on the: - Compliance with SOR requirements - Project scope definitions - Completion of all critical studies required in the FEED - Risks at this stage (what technical risks will be passed on to the contractor?)

(Continued)

TABLE 6.2 (*Continued*)

Workshops for Reviewing Front End Loading

No.	Timing	Workshop Subject	Participants	Objectives
3	After completion of the FEED and draft of ITB	Review of contractual/financial/commercial terms of the ITB - Contract team/s of the company with the project management team and design unit	- Contract team/s - Project management team and design unit - Relevant stakeholders	To review and check for: - Potential conflict, mismatch, and gray areas between contractual and commercial terms and conditions with technical scope, requirements, and specifications. - Contracting strategies[b] - Biddability[c]

[a] The above-mentioned workshops are over and above the usual quality control measures done during the FEED by the FEED contractor.

[b] Contractual strategies need to be finalized at this stage. Contractual/commercial/financial terms and conditions are generally standard for oil companies, but need review for conflicts and mismatches.

In case of disputes, in most of the cases, ultimately the technical requirements govern. The contractual terms tend to be one-sided.

[c] It will be beneficial to have a biddability review of the ITB for complex, major, and giga projects by a third-party expert before it is issued out. She can be from within the owner company, if available. A contrarian view will be helpful, at this stage, to plug all the holes so that the ITB does not go wasted. For market to respond positively and to receive reasonably good bids, such an exercise is essential.

7

Overarching Issues

The purpose of this chapter is to present certain overarching issues that usually get little attention in a project and are more likely to be ignored. These are the gray areas that senior management and project managers would want to push aside rather than confront and solve at the right time. But they can prove disastrous later. They are as follows:

- Project organization, team, and leadership at owner's side.
- Emotional intelligence.
- Managing studies during the project.
- Interface engineering during FEED.
- Project risk assessment.
- Value of project data.
- Lessons learnt.

They are discussed below briefly.

7.1 Project Organization and Leadership

Nothing is more important in a project than the project organization, team, and leadership. Research has shown that owner side project and engineering management play a key role in the project. It behooves the owner to have the right project organization from early stages of the project. Integrated project teams with members from different stakeholders in the company have proven to be the best.

Owner companies tend to have several specialist departments. Senior management must ensure that they are represented in the project team. The best structure will be a core team with an 'on-call' membership of specialists. The integrated project team can meet at regular intervals during stages 1, 2, and 3 of the project with a carefully prepared agenda, and create and maintain a list of issues with details of who has raised it and who are all assigned the task of resolving it, and a summary of the resolution. Unresolved issues remaining over a period of time must be flagged and raised to senior management.

DOI: 10.1201/9781003317081-9

Project and engineering management are complementary positions. If it is a major or mega project, both the positions could be parallel and reporting to the project director/senior executive. As with project managers, the engineering manager's position must be filled by a carefully chosen individual. For major and mega projects, the engineering manager's competency and background are very important. Just any middle-level engineer in the owner company will not fit the bill.

E. N. Merrow and N. Nandurdikar have written about the personality of project managers in their book 'Leading Complex Projects'. All their findings about successful project managers are true for engineering managers also.

They have analyzed the data on many projects and interviewed several project managers, based on the OCEAN personality model. The OCEAN model is a widely accepted model of personality by psychology researchers. The model basically distills the various personality factors into 5 types called the Big Five, namely, Openness, Conscientiousness, Extraversion, Agreeableness, and Neuroticism (OCEAN), and measures them using statistically validated scales.

Their finding is that successful project managers, though they are mostly like the cohort of engineers from where they come, are unusually open to new ideas and different approaches to problem-solving. They are generalists as project managers are, but are equally good at dealing with all types of work and people.

Furthermore, the finding is that for complex and mega projects, there needs to be true leadership, more than project management. The above applies to engineering management also.

Continuity of the owner side project and engineering professionals in a project is another finding that should be taken seriously by owner companies if they do not want projects to suffer.

7.2 Emotional Intelligence (EI)

Technical qualifications and competencies help engineers move up their career, up to a certain extent. Afterward, when it comes to leadership in the project and engineering management context, emotional intelligence is a key factor.

EI is factored into four essential elements, namely,

- Self-awareness.
- Self-management.
- Social awareness.
- Relationship management.

Each of them is important for a leader. There is an Emotional Intelligence Scale (EIS), which measures a person's EI. Like the OCEAN personality model, EIS has been used and statistically validated many times.

In their book, E. N. Merrow and N. Nandurdikar are unequivocal in their finding that successful project managers all have excellent soft skills and the majority of their time is spent on that dimension rather than on the technical side. In complex projects, leadership plays an important role and leadership requires a high level of EI. Owner company management must take note of the research findings while installing their project and engineering managers. The reader will be able to appreciate the role of engineering management when they read the sections on project risk assessment and interface management.

7.3 Managing Studies during the Project

Stakeholders in an owner company usually want several studies to be done before finalizing the FEED, and as a result, the engineering manager often gets stuck in the cycle of the ongoing studies, its reviews, and revisions of the documents.

At the outset, whenever studies are requested in the Statement of Requirements (SOR), the FEED team must review the same and understand the difference between a research study and an engineering study. A research study on an analytical side of upstream oil and gas requires collecting samples of oil, gas, and formation water and doing laboratory studies sometimes in another country. The same is applicable to the mixing of the effluent water from various reservoirs or for oils of different types, for example, heavy oil with light oils.

The second type of studies will be on selection of equipment. FEED often gets delayed because of studies put in for selection of equipment, for example, selection of the compressor type. Such studies on selection require direct feedback from actual users to be meaningful. A desktop study is not enough. All the above takes time. The problem is that the FEED schedule does not recognize these ground realities and does not provide enough time. In some cases, due to paucity of time, the FEED team decides to limit it to a desktop study, with dire consequences later.

Engineering studies should have definitive data available. The conclusions, based on either results of computer simulations or OEM inputs, should be actionable on the FEED technical documentation. The engineering manager must be alert to the results of such studies because it can go either way, meaning it can result in postponing or shelving of the project.

The other set of studies are from Health, Safety, and Environmental (HSE) aspects of the project. The number of such studies can only go up with the passage of time. Problems can arise if the owner company does not have guideline documents, systems, standards, or procedures to handle HSE studies. There can be endless arguments between the consultants and stakeholders on various assumptions and results of these studies. Therefore, it would help if each of the HSE studies be done on the basis of the Terms of Reference and Assumptions register, which should be carefully finalized beforehand. Engineering managers should ensure that adequate time is provided for all HSE studies and that mechanisms and procedures exist for the timely incorporation of the comments from such studies. The author (GUK) has seen projects where the HSE studies have delayed the FEED.

The other part is sometimes about who will pay for the HSE studies, the owner, or the FEED contractor? HSE studies are to ensure the owner's interest and should be paid by the owner. It will be best for the owner company to maintain a centralized team to manage HSE studies.

Lastly, it must be borne in mind that HSE studies such as Quantitative Risk Assessment (QRA), Safety Integrity Levels (SIL), Fire and Explosion Risk Assessment (FERA), though quantitative, are based on several critical assumptions. It is easy to get into a cycle of jargon-filled technical arguments on the above. The suggested solution is for the owner company to have a panel of world-class experts on a demand consultancy basis, who will be able to provide a way out of this 'analysis paralysis'.

7.4 Interface Engineering

Proper interface engineering during the FEED requires time and effort, and the engineering manager is accountable for the same. Interface is more than just hardware tie-ins and tie-in lists. While the Piping, Electrical, and Instrumentation disciplines usually develop comprehensive tie-in lists with details, what is missed often are the software and procedural interfaces with several entities and outside agencies. Plant operational data are routed to owners' existing systems, which could be old. Software/software version incompatibility is another issue to watch out for. If there are any procedures required, which call for reporting of information on new plant shutdown, flaring, etc., to related outside agencies (e.g., a downstream consumer of the oil and gas, government authorities), the same should be firmed up as a protocol with such agencies, to avoid confusion later.

When dealing with complex-mega projects with several work packages and contractors, an organizational interface diagram and register can be thought of, which will show the high-level interfaces between the work packages and the entities on both sides of the interface.

An interface register or still better, a database should be maintained and should be periodically reviewed by the engineering manager for updates from the field till the ITB is issued out. An interface diagram should be included for each interface. It will be beneficial to have the last update before finalization of the contract, in order to capture as far as possible any changes in the field. Many references and literature are available on the subject for a better understanding of the subject. For major or mega-complex projects, simple spreadsheets will not be enough. A suitable application software is recommended. It will help improve the efficiency of interface management.

7.5 Project Risk Assessment

We live in a VUCA (volatile, uncertain, complex, and ambiguous) world. With the pandemic and geopolitical situation, the resources are uncertain, supply chains are unreliable, money supply is restricted, and project decisions are more likely to be wrong than right. Project risk assessment is therefore all the more important.

While the project risk assessment will look at several risks to the project, the engineering manager will be focusing on the technical risks, namely, scope definitions, stakeholders' requirements, estimates/assumptions, specifications, procedures, technology/equipment selection, delivery, operational reliability, etc. The gray areas and unknowns need to be identified and recorded rather than being swept under the carpet. The engineering manager should have a clear idea of the technical risks that are being passed on to the contractor.

The author's (GUK) experience is that project risk assessment should not be done by the FEED contractor or the owner's project or engineering team involved in the project. If they do it, the objectivity will be lost and the risk assessment results tend to get white-washed. It is best done by a third-party expert outside the project. The expert can be inhouse or outsourced.

Some of the specific risks (apart from the usual technical risks in an oil and gas facility) to identify and handle are as follows:

1. Complex project design criteria.
2. Scope added in stages or scope creep since giving the SOR or mid-way through the FEED.
3. New technology/equipment untested in the field. (Unless a new technology/equipment is tested at the specific field conditions, accepting the same will be a risk.)

   Such new technology can be tested separately as a pilot project, but not as part of a project FEED with a definitive schedule and budget.
4. Very stringent outlet specifications limiting the technology to one or two Original Equipment Manufacturers (OEMs).
5. Too many interfaces, especially with outside agencies.
6. Complicated control systems.

7. Emergency shutdown and depressurization and associated flare systems of a large facility. (There is no need to wait for the contractor to find out that it is impractical to depressurize the entire facility at one time!) Review the situation during the FEED itself and include in the FEED package guidelines for the contractor.

8. Requests from stakeholders that are not engineerable practically. (For example, theoretically, it is possible to have a cannon built for sending a capsule by a moon-shot, but it is not safe or feasible practically with the current state of technology!)

9. Wrong packaging of scope. All of the maintenance works kept pending in a facility for several years cannot be packaged into an EPC contract. Such scope requires extensive field engineering and site investigations. Engineering of such scope is best done inhouse or outsourced and executed on a cost plus or re-measurable basis.

    Also, watch out for the inclusion of items not related to the main project objectives and scope. For example, if the main project scope is for a new long-distance pipeline, it would not be prudent to include the replacement of equipment inside a facility away from the pipeline route in the contractor's scope.

10. Overdose of chemical injections. Chemicals that are injected into oil and gas process fluids are generally corrosion inhibitor, scale inhibitor, emulsifiers, biocide, etc. The chemical injection points in a facility are usually decided by the concerned metallurgy/corrosion specialist in an owner company. Unless it is based on data and validated corrosion models, the number of injection points is usually taken from an existing similar operating plant. But for a new project, the inlet feeds streams and compositions, and therefore, its properties could be different. Chemical injections are required, but they tend to be costly and an optimization effort will be worthwhile in saving operational costs of the plant. Compatibility issues can also come up.

11. Compliance requirements with owner company standards/procedures/practices. The risks could be due to revisions of such documents during the ITB and bidding process, stakeholders' demands on retrospective compliance requirements, conflicts with international standards, out-of-date owner company standards/procedures, and conflicts with package vendors' standardized equipment standards.

12. Low capital cost with high operating cost. When considering equipment such as high-capacity compressors with high capital cost, there is a risk of going in for the lowest bidder (capital cost) with high operations and maintenance costs of equipment, which is often hidden. In such cases, a proper Life Cycle costing (LCC) will help in deciding between two options, all other factors being equal. The same needs to be clearly specified in the FEED with the LCC calculation methods.

7.6 The Importance of FEED Data in Oil & Gas Projects

During the authors' long stint in upstream oil and gas design/engineering and execution, we have come across several projects where the compositions of inlet FEED were not appropriate, incomplete, or plain simple erroneous. The ironic part is that all of these were avoidable!

7.6.1 Summary of Oil and Gas Process

A summary of the oil and gas processing fundamentals is given here, without details, so that the reader can appreciate the case studies that follow.

The reservoir fluid from the oil well passes through several pressure vessels in sequence as given in Figure 7.1.

The main equipment in the oil and gas processing lineup is usually the following:

A brief description of the Figure 7.1 is given below:

A. Oil side:

High pressure (HP) 2 phase (gas, oil, and liquid) separator.

Medium pressure (MP) 2 or 3 phase (gas, oil, and water) separator.

Low pressure (LP) 2 or 3 phase (gas, oil, and water) separator.

Crude oil stabilization unit (if required).

Storage tank/s of crude oil.

Desalter train/s depending on the need to dehydrate and desalt the crude oil.

Export pumps to the pipeline.

Note: The HP, MP, and LP separators are lined up sequentially and known as the separator train.

B. Gas side:

Compressor train/s HP/LP/MP. The gas is fed into the appropriate stage, based on its pressure.

Gas sweetening unit/s – depending on the need to remove hydrogen sulfide (H_2S).

Gas dehydration unit/s.

C. Effluent water side:

Effluent water treatment units: primary, secondary, and tertiary depending on the quality of the effluent water requirements for injection or disposal.

D. Others:

There will also be the associated safety systems, instrumentation and control systems, and utility systems.

Utilities and supporting systems, as well as buildings and infrastructure, will be included in the project depending on the facility design.

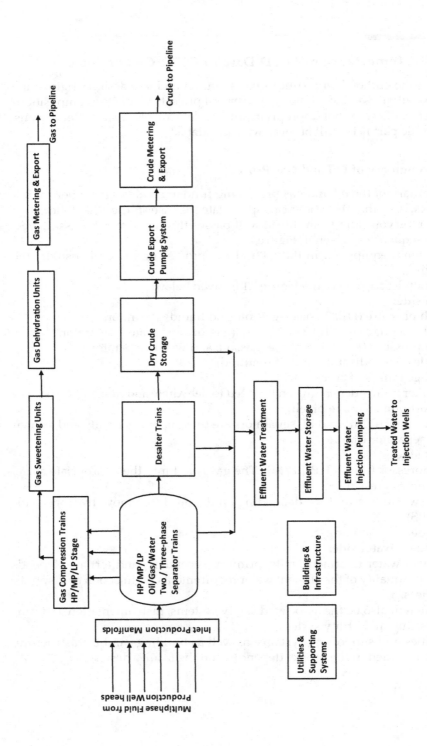

FIGURE 7.1
Typical oil and gas production schematic – simplified.

7.6.2 Oil and Gas Separation Process

The reservoir fluid is flashed in stages in the separator vessels, where it separates into hydrocarbon gas and liquid if it is a two-phase separator, or into gas, oil, and effluent water if it is a three-phase separator. The composition of the outlet steams, and quantity depend on the pressure and temperature at which it is flashed. The number of stages, high pressure (HP), medium pressure (MP), or low pressure (LP), and its temperature or pressures are determined using computer simulations. The simulators are compositional simulators, and the results totally depend on the inlet FEED compositions and conditions. The optimization of the conditions calls for several simulations and reviews. Obviously, the more representative samples that are available, the better will be the simulation results.

The optimization targets are the composition and pressure of the gas and the separated oil. Optimization of temperature and pressure will minimize the compressor cost and gas treatment units and stabilize the crude so that the oil has enough lighter components, which makes it more saleable, but to the right extent so as not to overshoot the pipeline transportation and shipping specifications.

The samples are obtained in several ways, such as bottom-hole samples and surface samples. The analyses also are done in several methods to ascertain the composition, as well as the properties of crude oil.

Crude compositions vary from C1, methane with a single carbon atom, to C60 (known as fullerene with 60 carbon atoms). When the subsurface/ reservoir teams report the analysis, it is usually up to a certain number of carbon atoms, say C20, and the rest are lumped together as C20+, known as the hypothetical component. The properties of C20+ are given, generally specific gravity and molecular weight, to make it complete. When specific gravity and molecular weight are given, the simulators can calculate the rest of properties and use it to solve the simulation model equations. Without the C20+ specific gravity and molecular weight, the compositional data are incomplete and cannot be used as input to the simulator.

The samples' composition is reported as 'mole percent' or 'mole fraction' on a dry basis. The percentage of water is given as 'water cut', which is the percentage of water in the total reservoir fluid. Therefore, to include the water content as mole percent, the dry composition is simulated with the required water cut values. The output stream of the simulations will contain the full composition of crude oil including water.

7.6.3 Compositional Data in FEED

Situations in FEED compositional data summarized below illustrate the different situations.

a. Upstream Facility Design with Limited Samples
There have been cases where entire oil gas facilities were designed with just one or two limited sample analyses. The problem with such

compositions is that they often do not represent the real nature of the reservoir fluid. Furthermore, it so happens that by the time the project gets commissioned, many years (5–7 years) would have passed. Reservoir fluid compositions from different wells could change, especially water cut and gas-oil ratio (GOR). While taking the samples, the ambient conditions tend to be important. For example, the samples taken during summer, winter, and other months would vary in composition within a certain range.

The author (GUK) had come across a case of a high-capacity oil and processing facility that was designed and constructed based on just two samples from a couple of wells. As per the report and recommendations from the subsurface teams, there was also a separate train to process a special crude containing a high percentage of asphaltenes. The project took about 5 years to complete. After start-up and commissioning, it was found that the entire gas side has been overdesigned including gas compressor, gas dehydrating unit, and related piping and control systems. Considerable time and efforts were spent on modifying the control system and its settings to make the system work. The gas side went onstream eventually but was not operating in stable conditions. It was operators' nightmare with frequent upsets, shutdowns, and flaring of the gases!

And the special type of oil with high asphaltenes never materialized! There was no crude feed even to commission that special train. It remained idle for a long time. The train had to be modified considerably to suit a different type of crude.

b. Incomplete or Inadequate Compositions

As mentioned in the previous illustration, often the composition is given without the necessary data to compute the properties of the hypothetical parameters such as C10+. Even if the contractor requests for the data, it may be difficult to give, simply because they are not available. The old sample cannot be recreated in the present. And any new sample composition can be different. There was an instance where the owner company could not provide any further data. The contractor was left high and dry. They used all their wits and expertise on the simulation software to produce the properties of the hypotheticals. Obviously, it did not match with respect to the reality. This brings out another issue of verification of the compositions, which is given in a subsequent section.

In case of the compositions, it is always recommended to do the analysis of the components as high as possible, in the ranges of C30 to C40. The method of analysis used, such as chromatography must be mentioned in the report.

c. Abundance of Compositions

Another issue the authors have seen is the abundance of sets of compositions. The owner company to protect its part throws into the ITB technical documentation whatever compositions are available from all the wells that could be connected to the oil and gas facility, without informing the contractor on what basis or proportion the streams will have to be mixed and fed into the facility. Of course, computer simulations can be performed a million times, but what will be the basis on which the process equipment will be sized?

There was a project where feed streams from four fields were connected to the facility. Let us call the fields A, B, C, and D. There were ten wells each from a field. Clearly, there were different combinations as to how the streams (flowlines) from the well could be opened up to the facility. Three of such connections are shown in Figure 7.2. All flowlines are connected, but all the valves are not opened simultaneously to make the fluid flow go into the facility. The valves shown in black are closed. The facility has a certain capacity for processing crude oil, gas, and water, and the number of flowlines connected to a facility is limited by the processing capacities of the facility. The connectivity is available, but all of the wells cannot be opened up to the facility!

When the contractor requested for the combinations for use in sizing the equipment, there were confusion and delay in responding to the contractor's request. Owner company's subsurface and production teams could not readily agree on the values. The contractor was asked to submit different combinations for review. And when the owner thought of finalizing one particular combination, it was found that such a combination produced higher quantities of gas, and therefore, the size of the gas compressor will be higher. The contractor promptly declined to use the combination, stating that the gas compressor size and its costs will be much higher than what they had bid for the work. Moreover, the ITB did not include anything specific about the combination to be used by the contractor and that they are not bound contractually.

d. **Change in Input Feed Compositions after Signing the EPC Contract**
One of the major impacts of a project is changing the compositions after signing the EPC contract. Normally, the owner companies accept the situation as a change to the contract and negotiate with the contractor for cost and schedule impact. But such situations are fraught with contractual and legal implications. Audits tend to view the above occurrences suspiciously.

There is a chance that one of the original bidders to the ITB, especially the next lowest, raises the point of a major change soon after signing of the contract, as a violation to the bidding norms. Mega project contracts are quoted in a very competitive basis, and so all bidders are alert to the situation.

Needless to mention, the above situation is to be avoided as far as possible.

e. **When Contractor Finds Changes to the Compositions**
Another case that the author (GUK) has experienced is the requirement placed in the contract for the contractor to carry out the crude stream

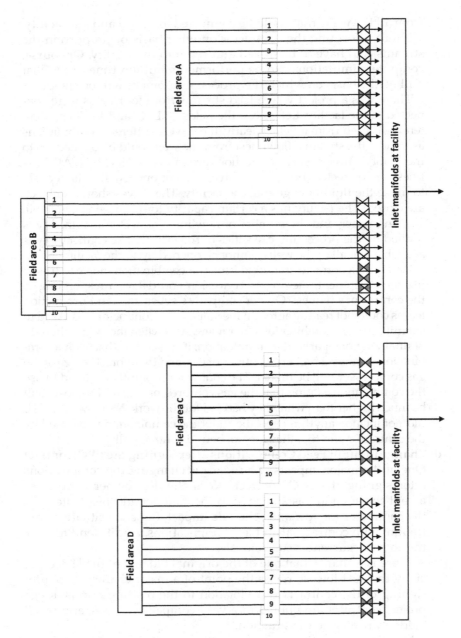

FIGURE 7.2
Connectivity of flowlines from fields A, B, C, and D.

analysis and reconfirm the compositions. It so happened that when the contractor did the analysis, the composition came out to be different, with the potential impact on the compressor operating range.

The contractor requested guidance from the owner company. Naturally, the concerned departments in the owner company burned midnight oil to come up with a solution. Due to related delays, the contractor had to be given an extension of time and certain cost adjustments.

f. Soil Investigation, Geotechnical Data, and Site Data

There are several major and giga projects that are replete with rain of change orders at the start of the project due to revisions in the soil investigation and geotechnical data provided in the FEED, by the owner.

The usual practice from the owner side is that during the start of FEED, the soil investigation and geotechnical data work is given to the concerned department or outsourced. The information in the report received from the outsourced agency is used for the preliminary design of the foundations and structures. It is included in the FEED/ITB. The quality of the report from the above must be checked rigorously to avoid mishaps in the future. Most companies have a checklist to verify the quality of the reports on the above. If such a document is not available, it must be developed and put in place.

During the bidding process, there is practically no time for the bidder to verify the data provided by the owner. It is taken in good faith.

Later when the contractor does the soil investigation and if the data are found to be at variance, problems crop up.

Therefore, the owner must do a thorough check on the data, especially if the project site is in problematic areas such as sand, marsh, 'sabkha', subsea, landfill. Large storage tanks in the above areas are particularly vulnerable. Ground Penetrating Radar (GPR) survey must be used to detect underground services.

Site data such as climatological data are usually developed once and updated periodically by the owner companies. Climate change and seismic zones must be updated based on the latest data from research. For example, the severity of extreme weather (extremes of temperatures, storms, floods, etc.) has increased in several areas in recent years. Data that are to be used for the design of a facility that will be in existence for another 15 to 20 years in the future must consider the impacts of the above.

Important points to note:

1. Process equipment are not smart like human babies that grow and learn as time passes. They can do the operational functions it is designed to do within a certain window, called the operational window. It can do this without problems 365 days in a year. Each equipment has an upper and lower design limits for the process parameters applicable to it, and the control system takes care of the movements within the window (control range) smoothly. If any of

the parameters exceed the operational window, process upsets, incidents, and accidents can happen. The entire subject of process safety has developed on the above concept of protecting the system and subsystems from abnormal operations.

2. Specifying inlet feed compositions is important, and the owner has to spend time and effort to develop the process models and do the required simulation cases.

3. The design/engineering section within the owner company usually does not have any expertise on subsurface, and they do not understand or question the analysis that comes from the subsurface teams. The design/engineering unit should acquire certain expertise in reservoir engineering to properly understand the terminology of the subsurface terms and its implications.

4. Compositions must be sufficient and should include the technical information for input to the simulation model. Whenever compositions are given, the owner company must specify how the several input streams and compositions have to be mixed and fed into the facility. If streams from a couple of wells from a particular reservoir contain high GOR, how it has to be mixed with other streams? If the high GOR streams are included, naturally the gas quantities and qualities will be impacted. In case of water cut, also the situation is similar.

5. The owner company's subsurface teams must also do forecasting in order to estimate how the pressures, temperatures, water cuts, and gas-oil ratios are likely to change in the future.

6. Ideally, it will be better to have a document signed by the concerned stakeholders in the owner company regarding the input data and compositions that will be used in the project. The data to be used in a major /mega project need to be discussed with the design unit by the subsurface teams to understand the potential impact on the surface facility design. There must also be a clear understanding that any updates regarding the data during the course of the project duration will be shared and discussed with the design and project teams to study its impact on the same. Changes in the data should not be swept under the carpet. A plan for how to handle changes in input data must be included in the document mentioned above.

7. Project governance systems and procedures should include the above scenario as part of its Management of Change (MOC). The subject of FEED parameter change does not appear in most of the Project Gate System documents that the author (GUK) has seen.

8. Soil investigation and geotechnical data in the FEED require careful attention and quality checks. It will be best if the owner

company keeps track of the data given by the company in various projects, as well as those of the contractor, and analyzes it periodically to see if there is any variance and, if at all, from where it has come. The areas where the most variance occurs can be flagged for further investigation and capture of better data. A central depository will be required for handling the above, and it will be worth the effort.

7.7 Value of Project Data

Earlier sections discussed about the importance of inlet data in a project, whereas this section highlights the importance of project data or 'as-built' information that is available once the project is completed.

There are two instances in the life of a project where valuable information is available: one at the end of FEED completion by the FEED contractor and the other at the end of project completion at the EPC/LSTK contractor.

FEED documentation handover in owner companies is sometimes hazy, because unless there is a central depository, who owns the FEED data is a question mark. The authors have seen owner companies in the paper era who were managing the mountains of paper documentation quite efficiently. With the advent of digital storage and retrieval, it should have been easier, but on the contrary, it has become difficult.

The key issues here are the following:

a. Who will own the data? Traditionally, the central engineering departments were in charge of the above. But with downsizing and resource crunches, the owner companies are lax on FEED documentation. The FEED documentation is usually with the FEED contractor and will be handed over to the owner project teams when their work is over and is kept somewhere (mostly in the records room or-library sections, if at all they exist) with no ease of access and retrieval. The project proponent sometimes gets a copy.

b. Lack of systems and procedures about how to store data and its retrieval mechanisms. Paper documentation is definitely not feasible. Owner companies need to spend time and effort to decide about the software for a comprehensive document management system to be used for the above. Its compatibility with the FEED documentation must be checked and demonstrated. For example, facility drawings will be in 3D software and they must be readable (if not editable) with the retrieval software. The same is applicable to all documentation. Query, search, and retrieval setup must be relevant to the context of the FEED. Of course, resources must be available or outsourced. (For example, all the FEED cannot be in a single file. It must be classified and coded separately so that the query can bring up the relevant files).

c. FEED data become valuable in the next similar project or making modifications in the facility. Granted that as-built data from the project are more easily available, however, only the full FEED information will give answers to why and how the facility components are designed that way.

The second set of information is from the EPC contractor, who, in order to fulfill the contract requirements, prepares the 'as-built' data and hands over the project information to the client. Again, truckloads of box files containing the as-built documentation are definitely out. But digital as-built is seldom checked for compatibility.

The sections a, b, and c noted above are applicable here also. The ownership issue is better here, because the operating entity of the owner company takes care of the operating and maintenance manuals.

Detailed, clear, and precise instruction must be included in the contract as to how to prepare the as-built documentation and its delivery. (Stating 'soft copy' or in a 'certain format' will not help).

The following cases are examples of problems faced with as-built documentation:

- The 'as-built' is only a one-to-one digital copy of the old paper style as built, because there is no clear instruction from the owner company on how to prepare the same.
- Submitted as-built documentation in CDs could not be opened.
- The entire drawings are in a single file!
- Drawings in individual files, but with drawing numbers only, no titles, which makes search difficult.
- No standardization between projects and between other stakeholders.

A good, well-maintained digital document management system can save owner companies', time, effort, and money that they end up spending when they embark on new projects or modification of existing facilities.

7.8 Lessons Learnt

The peculiar finding about project lessons learnt is that a simple internet search will reveal wealth of literature on the subject, that too with reference to oil and gas projects. But the authors' experience is that in reality, very few companies have lessons learnt data that is really useful.

We think the root of the problem is that the people involved tend to think of lessons learnt in terms of paper and maximum in terms of spreadsheet. The above thinking will only translate data into spreadsheets or certain other limited applications. What is needed is to think in terms of knowledge management and software that can capture the data into a true database that contains the relevant information and has ease of retrievability. The above should be part of the permanent overall project knowledge management depository system of the company with access permissions and outside the temporary project management organizations. Rather than trying to develop such a system on their own, it will be worthwhile to bring in a knowledge management specialist consultant to be embedded within the project teams and develop correct and useful systems.

Project and engineering management must carefully define the terms and discuss with the stakeholders regarding what constitutes lessons learned. Lessons learnt is not a blame game, and should be finalized after meeting or consultation with stakeholders. There is a wealth of literature available on the subject. Again, third-party involvement may be required as stakeholders tend to avoid reporting misses from their side.

Part II

Case Studies and Lessons

8

Case Studies: What They Tell Us

The following chapters present case studies that illustrate how the gaps, errors, and mismatches in the FEED impacted projects as they progressed. As can be seen in retrospect, all of them were avoidable. The project identities have been masked for confidentiality.

There is a serious deficiency in owner companies about collecting and keeping such case studies. Secrecy prevails due to several reasons, the blame game being the foremost. The authors' humble opinion is that the owner companies collect such case studies and build a database of the same so that it can be used to avoid similar issues in the future. An organization-wide attitudinal change will be required to minimize design errors.

There are databases of accidents available in several sectors including the chemical and hydrocarbon industry, some of which are public and some paid. Though many papers have been published on design errors, there are no large-scale databases available on design errors in the process industry. Academic research needs to pursue this aspect.

The shedding of engineering expertise from owner oil and gas companies is a common factor that stands out in all case studies. Owner companies must think of expertise as intellectual capital and national asset. The overheads of keeping them on company payrolls are little compared to the impact they can make in capital projects. The thought that we can save costs as design and engineering is not our core business and can be outsourced is a short-term sunrise-to-sunset view, which has spread in owner companies. It has done enough damage. Both authors have experienced the 'bewilderment' in owner companies when problems come up during FEED and later in detailed engineering and construction. When the ball lands in the owners' court, they find that they have no resources to provide any answers that are in their interest.

The authors are not alone in the above findings. Several consultants and researchers have arrived at the same conclusion. It merits serious thought among senior management in oil and gas companies.

9

The Baywater OPCU Project – A Risk Not Worth Taking

This is a case study of a project where the client and the contractor took on certain risks, which ultimately resulted in a project that has stalled, with the client and the contractor unable to resolve the issues that stalled the project.

Project Overview:

Project Name – The Baywater OPCU Project

    Client – The national hydrocarbon authority of a South Asian country

    Contractor – International Marine Project Ventures, Singapore

    Project Award Date – June 2015

    Project Value – USD 150 Million

    Current Project Status – Originally contracted for completion by June 2017, currently stalled at around halfway to completion, with client and contractor unable to reach an agreement for resolution.

    Project Description: The National Hydrocarbon Authority (NHA) is the government authority tasked with hydrocarbon energy exploration, production, and distribution, in a South Asian country.

NHA has initiated redevelopment of an offshore gas field, which has been in production for the last 5 years, and where some new reservoir capacity has been recently identified.

7 wellheads are currently in production in this field, and the redevelopment plan envisages installation of an additional 5 wellheads and a new gas export pipeline leading to a new landfall location.

The drilling, wellhead installation, and the new subsea pipeline works are in progress, and the NHA is looking at creating a new mobile offshore gas processing and compression unit (OPCU), which will be the last link in the gas production and export chain.

Based on Feasibility studies carried out by their inhouse engineering team, the NHA has identified that one of their existing mobile offshore drilling units, which is currently 'mothballed', can be converted to the OPCU as required to complete their project.

The NHA had issued a tender for converting the offshore drilling unit to the OPCU.

DOI: 10.1201/9781003317081-12

Project Scope: Design and fabricate new topside process modules consisting of the following:

- (PGC) Process Gas Compressor (2 trains×1.8 MMSCMD each).
- GTG (Gas Turbine Generator) (2 trains×3 MW each).
- Knock-out drum (3.6 MMSCMD & 8000 BLPD of liquid).
- Flare tips.
- Produced Water Conditioning System (2 hydrocyclones×50% & 1 Degasser Unit×100%).
- Chemical injection, N2 generator.
- Hydrochlorination, flare boom.
- Utilities.

Upgrade and refurbish the following existing systems/equipment:

- Jacking system.
- Legs.
- Helideck.
- Bilge pumps.
- Diesel purifier.
- Air compressors.
- Rescue boat.
- Demolition of existing equipment.
- Integration of new equipment at the fabrication yard.
- Marine transportation and installation.
- Offshore hook-up, pre-commissioning, and commissioning.

The Contract Award: The NHA awards an EPC contract for the conversion project to International Marine Project Ventures (IMPV).

On a detailed review of the project scope, IMPV determines that it needs to bring in a JV partner to plug certain competency gaps in IMPV's current operations. IMPV chooses ABC Shipyards (ABCS) as a JV partner, with the specific scope of undertaking the upgrade and refurbishment, demolition, marine transport, and installation scope.

The NHA agrees to the JV approach with the condition that IMPV shall be identified as the prime contractor.

ABCS in turn engages a specialist engineering consultant, Marine Engineering Services (MES), for design and engineering services related to upgrade and refurbishment.

The overall contract is valued at USD 150 million, with a 60%/40% scope split between IMPV and ABCS.

The contractor's pre-award checklist is given in Table 9.1.

One important conclusion that we can draw from this risk assessment is that the contractor has assumed that the uncertainties with respect to the condition of the existing systems and facilities, and the feasibility of their adequate refurbishment for the new service, carry a low risk and can be easily managed.

However, when we look at the status of the project after 24 months into construction, the cascading effects of delays in detailed engineering, procurement, and fabrication can be seen as shown in Table 9.2.

TABLE 9.1

Pre-award Risk Assessment by Contractor

Sl. No.	Potential Risk Identified	Risk Rating	Mitigation Actions
1	Extremely tight overall schedule – engineering and yard	High	1) Advance engineering 2) Pre-bid tie-up with GTG and PGC vendor
2	Jacking System Adequacy requiring Modification of Jacking System	Low	Topside equipment selection to limit the weight
3	Jacking System Adequacy – ambiguous scope for gearbox refurbishment	Low	Cost for the pessimistic scenario
4	Lengthy flare boom and increase in weight for flare and supporting weight	Medium	NHA has agreed to consider the staggered release of hydrocarbons during emergency shutdown to reduce flare boom length
5	Refurbishment of legs (rack, chord, and tubulars)	Low	Based on past experience, consider replacement of tubulars only Advance engineering and procure higher thickness tubular Replace all the braces in underwater leg sections
6	Possibility of inadequate draft during the MOPU redelivery	Low	Dredging at yard
7	Stick-built approach alongside jetty for process equipment	High	Augment material handling capacities on a rental basis and deploy experts from MFF Hazira

TABLE 9.2

Project Status 24 Months After Award of Contract

	Description	Weight Factor (%)	Contribution to Overall Progress Sch. %	Act. %
1	Detailed engineering	3.5	100.00	74.88
2	Procurement	55.0	100.00	61.70
3	Demolition	2.0	100.00	100.00
4	Fabrication	21.5	100.00	30.36
5	Transportation	6.0	100.00	50.00
6	Jack-up	1.0	100.00	0.00
7	Hook-up, testing, and pre-comm.	10.0	100.00	0.00
8	Survey	1.0	100.00	40.00
	OVERALL PROJECT PROGRESS	100.00	100.00	48.49
	IMPV PROGRESS	100.00	100.00	70.60
	ABCS PROGRESS	100.00	100.00	15.33

What Went Wrong: The key issue that stalled the project was that the refurbishment of the platform jacking system, which is a critical component of the final assembly, was not found to be technically feasible. The contractor's original assumption was that this was feasible without a major cost impact, but once the final topside design was completed, it became clear that the existing system was fundamentally inadequate, and a completely new system was required. The cost of a new system was much higher than originally estimated and would lead to a substantial loss for the contractor. Negotiations between the client and the contractor for a contract price revision have been deadlocked.

KEY TAKEAWAYS

1. In an EPC project where there are a significant number of unknown factors, which can impact the schedule and cost of the project, typically as is found in projects involving refurbishing or revamping an existing facility, the project owner needs to carry out a thorough engineering study of the existing facilities to clearly identify their condition and limitations. Passing on this responsibility to an EPC contractor is extremely risky because the EPC contractor has already committed to a fixed cost and a time schedule for the project, and their subsequent evaluation of the unknown factors could substantially jeopardize their cost and time commitment.

2. EPC contractors generally tend to assume a best-case scenario when faced with the problem of dealing with unknown factors at the pre-bid stage, because of the pressure to come up with a competitive bid, and the lack of time and opportunity to carry out a detailed evaluation of the unknown factors.

3. An early investment by the project owner in carrying out a detailed engineering study of his existing facilities, prior to putting out an EPC tender, could have resulted in a higher initial investment for the project, but would have ensured timely completion of the project and its subsequent operational benefits.

10

When Pipe Thickness Went Wrong

Project overview: The project involved long-distance pipelines carrying crude oil from one point to another. There were several pipelines with diameters ranging from 8 to 24 inches. The pipelines were partly aboveground and partly underground. The distances traversed by the pipelines varied from 20 to 50 km. The underground portion was longer in all pipelines.

Name: Pipeline project ABC

Client (owner company): Major oil company (OWN1)

FEED & Project Management Contractor: International engineering company (ENG1)

Contractor: International EPC contractor (EPC1)

Project location: Asia

Project duration: 30 months

Project value: Between 700 million and 800 million USD

Summary of project scope and description: EPC contractor scope included detailed engineering, procurement, installation, testing, commissioning, and handover of several pipelines to the client. The pipeline diameter varied from 8 inches to 24 inches. Pipelines traversed aboveground (minor portion) and underground (major portion). Lengths varied from 20 to 50 km.

Bidding and contract award: The Invitation to Bid (ITB) package with FEED technical documentation was prepared by an international engineering company (ENG1). ITB was sent to shortlisted EPC companies. Out of 4 bids received, L1 (the lowest) bid was from a well-known EPC contractor (EPC1). Curiously, there was a considerable difference between the L1 bid value and the next highest L2 bid (in the range of 4 million to 6 million USD).

After the usual rounds of bid clarifications, the bid was awarded to EPC1, whose bid was accepted as technically suitable and the lowest.

What went wrong: Within the first 3 months of the contract award, the contractor EPC1 came up with a formal request for using lower thicknesses in most of the pipelines as compared to the pipeline thicknesses given in the FEED documentation. The formal request was backed up with extensive calculations of pipe thicknesses, which also were submitted to the owner company.

The owner company's engineering contractor ENG1 duly reviewed the above calculations and found them correct. The above finding raised the following critical questions:

DOI: 10.1201/9781003317081-13

1. Correctness of the original thicknesses given in the bid (FEED) and in the contract award, which were higher for most of the pipelines.
2. About several contractual issues as to if the FEED thicknesses are minimum and need to be complied with.

Apparently, if the EPC contractor is right, then the FEED thicknesses were an overdesign and could be in error. And if the owner agrees to lower thicknesses, EPC1 will save a good amount of money. Can the savings be passed on to the owner company, and if so to what extent? All the above become issues when viewed from a contractual angle, since the bid, quotes, and the award were based on higher thicknesses in the FEED.

Analysis: Due to the complex nature of the issue and criticality of resolving the same as early as possible, an independent Task Force was constituted by the management of the owner company. The members of the Task Force were from outside of the engineering contractor ENG1 and the project team of the owner company. The time given to the Task Force was about 3 weeks.

The Task Force went through the FEED calculations, FEED thicknesses, and compared the same with the calculations submitted by the contractor EPC1. Connected clauses in the contract were also examined.

The findings of the Task Force are summarized as follows:

1. The calculations submitted by the contractor EPC1 were technically correct.
2. The FEED thicknesses were an overdesign, resulting from taking wrong temperatures of the earth for the underground portion of the pipelines.
3. Contractually, there were inconsistencies in the terminology used. In one instance, the thicknesses were stated as 'preliminary' and the responsibility was on the contractor to recalculate and confirm the thicknesses using various methods, whereas in another place, it is noted as 'minimum'.
4. The aspect of passing on the savings due to the efforts of the contractor to the owner company was not mentioned specifically in the contract. The contract was stated in general terms regarding the need for optimization by the contractor.

Following related pipeline thickness calculations can be safety skipped by the reader. Those who are interested can go through the same to get a deeper understanding of the issue.

Pipeline thickness calculations: An overview of the thickness calculations is given below without going into the finer details.

The forces in a pipeline under pressure are shown in Figure 10.1.

As can be seen, the pipelines are primarily under three forces. First is the hoop stress, because it is circumferential and tends to break the pipeline in the plane of the diameter; second is the axial stress since it will burst the

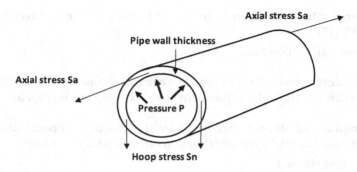

FIGURE 10.1
Forces in a pipeline under pressure.

pipeline along its length, that is, in the axis or longitudinal direction; and third is the radial stress, which is significant only in thick-walled pipelines. Transmission pipelines are considered thin-walled, and therefore, the radial stress is neglected. Usually, the dominating stress is the hoop stress, when the pipeline is not restrained, that is, not constrained by an outside mechanism.

On the contrary, when the pipeline is buried underground, it is restrained, and several other forces come into effect. They are discussed below with reference to **ASME B31.4 (Pipeline Transportation Systems for Liquids and Slurries)** and **ASME B31.8 (Gas Transmission and Distribution Piping Systems).**

The formula given below is in the English (FPS) system based on ASME 31.8.

$$t_{min} = \frac{P \times D}{2 \times S \times E \times F \times T} \tag{10.1}$$

$$t_{calculated} = t_{min} + CA \tag{10.2}$$

where

P – internal pipe design pressure, psig.

D – nominal pipe outside diameter, inches.

t – nominal pipe wall thickness, inches.

S – SMYS of pipe material, psig.

E – seam joint factor, 1.0 seamless and submerged arc welded (SAW) pipes.

F – design factor, usually 0.72 for liquid pipelines. For gas pipelines, the range is from 0.72 to 0.40 in class 4 locations. Class locations for gas pipelines depend on the population density in the vicinity of the pipeline.

T – temperature de-ration factor equal to 1.0 for temperatures below 250°F (121°C).

CA – corrosion allowance.

For buried and restrained pipelines, the t calculated thickness has to be cross-checked with the allowable equivalent stress coming on the pipeline from

- Longitudinal stress experienced by the pipeline – combination of effects of internal pressure (Poisson effect) and temperature.
- Bending stress.
- Axial stress – could be tensile or compressive.

Both ASME 31.8 for gas and ASME 31.4 for liquid pipelines require this check.

For the above, bending stress is usually negligible for straight length installed underground. The bending stress could be high during installation and is minimized by controlling the elastic bend of the pipeline. Here, it is taken as zero for simplifying the calculations.

Generally, no axial stress either tensile or compressive is applied to the pipeline externally. Therefore, it is also taken as zero here for simplifying the calculations.

Equation 10.3 gives the longitudinal stress S_L in mathematical terms.

$$S_L = S_P + S_T + S_X + S_B \qquad (10.3)$$

where

S_P – longitudinal stress due to internal pressure.

S_T – longitudinal stress due to thermal expansion in restrained pipelines.

S_X – axial stress, taken as zero here.

S_B – bending stress, taken as zero.

Going further,

$$S_P = 0.3 S_H \qquad (10.4)$$

$$S_T = E\alpha(T_1 - T_2) \qquad (10.5)$$

where

S_H – hoop stress calculated using the selected thickness, which will be $> t_{calculated}$.

E – modulus of elasticity of steel.

α – coefficient of thermal expansion of steel.

T_1 – installation, back-filled temperature (max or min).

T_2 – maximum or Min operating temperature.

As per ASME codes, aboveground pipelines are unrestrained, and therefore, for such sections, aboveground longitudinal stress need not be checked. But equivalent stress due to internal pressure S_P, as well as bending stress S_B, needs to be evaluated.

The limits of longitudinal stress established by the codes are given by the two equations:

$$S_L = S_P + S_T \leq 0.90.\text{SMYS} \tag{10.6}$$

$$\left(S_P^2 - S_L.S_H + S_H^2\right)^{1/2} \leq 0.90.\text{SMYS} \tag{10.7}$$

where

SMYS – Specified Minimum Yield Strength of the pipeline material.

Having gone through the above, let us see how the calculations were done in the FEED technical documentation for the pipeline design.

The FEED team used the calculation methodology correctly; firstly, they arrived at the thickness based on the hoop stress equation, added corrosion allowance, and then selected a suitable pipe thickness higher than the above.

Secondly, they did the cross-checking based on the equivalent stress method and evaluated all the terms in Equation 10.3 using the elaborate von Mises method. The cross-checking results showed that the thickness selected based on the first step did not satisfy the criteria as per Equations 10.6 and 10.7. Therefore, the thicknesses were increased to meet the criteria and a note was added to state the above in the pipeline design basis document.

The error happened while taking the installation temperature for the underground portion of the pipeline for use in Equation 10.5.

The owner company had a standard for climatological data, and the different temperatures to be taken for various calculations were indicated in the same. They are given below for illustration. The temperature values given are not real. For example:

- Aboveground temperatures from summer and winter average dry bulb temperatures:
    47°C/7°C.
- Underground temperatures 1.22 m below ground level from summer and winter:
    35°C/17°C.

Further, the pipeline design basis document stated the pipeline operating temperatures as 50°C max and 15.6°C min.

The FEED team while taking the temperatures for calculating longitudinal stress as per equation somehow used the temperature of 50°C and 15.6°C for T_1 and T_2 in Equation 10.5, instead of 35°C and 17°C.

The above error resulted in not meeting the criteria in Equations 10.6 and 10.7. The note in the FEED document clearly mentioned that the thicknesses have been increased to meet the criteria in Equations 10.6 and 10.7.

The impact resulted in unnecessary higher pipeline thicknesses effectively increasing the total pipeline tonnage by 20,000 tons!

Not to mention the contractual wrangling and project delays involved once the contractor arrived at the correct thicknesses, which were lower!

Authors' note: The baffling fact is that how the experienced FEED team missed the above. The key points in the issue are as follows:

- An experienced pipeline design engineer should have noticed something was amiss when the equivalent stress calculations indicated non-compliance to the criteria. Adjusting the thicknesses to meet the criteria is an unusual design practice for conventional crude oil lines.

- The departmental design review and overall FEED documentation review should have brought out the error, which points to the quality of such reviews. 'Check the box mentality' of the review team members may have contributed to the above. The review team members may have been exhausted. There could be several factors.

- It took only 3 weeks for the Task Force to identify the design error, and that too with team members who were not experts in pipeline design. In fact, the young piping engineer in the Task Force was new to the equivalent stress checking methods. This shows that whenever there is a clear target for the team, and the right leadership, design reviews can be very useful.

The annals of design companies are filled with such error stories!

The key takeaway is the need for independent review of all design documentation, at the discipline departmental level and overall FEED level. Ensuring the quality of such reviews is not easy in a mega project, but well-established techniques can help, which are given elsewhere in the book. Time must be available for such reviews. The engineering management leadership plays a key role in ensuring the quality of engineering deliverables.

11

When Compressor Selection Went Wrong

Project overview: The project involved replacing old reciprocating compressors (total of 6 numbers in sets of 2 for 3 locations) with suitable new machines. The machines were used to compress high molecular weight–associated gases from near atmospheric pressure to about 7 barg (about 105 psig).

Name: Compressor replacement project (CRP)

Client (owner company): Major oil company (OWN2)

FEED & Project Management Contractor: International engineering company (ENG2)

Contractor: International EPC contractor (EPC2)

Project location: Middle East

Project duration: 24 months

Project value: About 100 million USD

Summary of project scope & description: EPC contractor scope included detailed engineering, procurement, installation, testing, commissioning, and handover of six compressor units with all accessories to the owner. The type of compressor had been fixed in the FEED documents by the owner company OWN2, as a wet-type screw compressor.

Bidding and contract award: The Invitation to Bid (ITB) package with FEED technical documentation was prepared by an international engineering company (ENG2). There was a good response to the ITB from established contractors.

After bid clarifications, the bid was awarded to EPC2, whose bid was accepted as technically suitable and the lowest.

What went wrong: The project execution stage was completed without major hiccups with only a slight delay. When the first compressor unit was commissioned, there was the problem of frequent compressor shutdowns due to lube oil filter choking. The lube filter system was duly cleaned out, and the unit was started up again. But the problem persisted. The same issue surfaced during commissioning of the remaining compressors in other locations.

The contractor formally informed the owner about the problem and suggested analyzing the material that choked the lube oil filter, since they suspected the same to have originated from the gas coming from the owner's processing facilities.

Advanced chromatographic analysis was conducted on the chock material samples, by the contractor and the owner. Analysis results from both sides

DOI: 10.1201/9781003317081-14

TABLE 11.1

Gas Composition

Sl. No.	Component Name	Mole %
1	Nitrogen (N_2)	4.37
2	Carbon dioxide (CO_2)	1.34
3	Hydrogen sulfide (H_2S)	0.20
4	Methane (CH_4)	71.10
5	Ethane (C_2H_6)	5.30
6	Propane (C_3H_8)	5.60
7	n-Butane (nC_4H_{10})	2.48
9	i-Butane (iC_4H_{10})	1.41
10	n-Pentane	1.42
11	i-Pentane	1.90
12	Hexane (C_6H_{14})	1.40
13	C7+	3.48

indicated the presence of hydrocarbons in the chock material, particularly higher molecular weight hydrocarbons from C12 onward, indicating the presence of crude oil.

The above presented a mystery. The project remained unsuccessful till the mystery was solved.

Analysis: The following were the data with respect to the feed gases:

1. The gases were coming from the crude oil storage tank, which was the last step in the process of separation of crude oil, water, and gas. From the storage tank, the crude was pumped out for export. The tank was fixed cone roof tank.

2. Analysis of the gas itself did not reveal anything unusual. The typical gas composition looks like the one in Table 11.1.

The component composition does not remain steady throughout the year; it varies depending on the weather conditions, generally +/−3 to 5%. The gas is comparatively heavy with molecular weight in the range of 37–42.

The gas composition analysis above is on a 'dry basis'. Water is not included. To find out the water mole percent in the gas, the temperature and pressure conditions of the gas must be simulated to know how much water the gas can carry at those conditions.

Most importantly, the gas compositions given in the FEED and contract did not mention any liquid content in the gas.

The investigation committee set up to handle the issue interviewed the persons in the owner company who were involved in the data input to the FEED. The engineering company who prepared the FEED was no longer in the picture, and so only the records were available. The inputs from the personnel and the records did not show anything unusual, except for a recommendation report from the FEED contractor on selection of the compressor.

The document elaborated on the selection of the wet-type screw compressor. Wet-type screw compressors have the rotating screw bathed in lubricating oil, which is recovered downstream by a suitable lube oil filter and recycled. The lube oils used are not proprietary. The selection was based on the 0% to 100% turndown possible in the screw compressors and its insensitivity to changes in the gas molecular weight. The other main contender was 'dry-type' screw compressor, but the same was ruled out since the evaluators thought it has very finely machined parts/surfaces and requires costly maintenance. Centrifugal compressors were not favored due to the lower gas capacities required to handle and the machine's sensitivity to molecular weight changes.

The committee members noted the following in the report of the FEED contractor:

- There could be a potential problem of the gas reacting with lube oil at the higher temperatures of compression. The report recommended that the contractor/compressor manufacturer confirm the suitability of lube oil for the gas being handled in the compressor.
- The original recommendation report's conclusions on selection of the compressor type were vague and not firm about the selection. It had lots of 'ifs' and 'buts' and left the final decision to the owner company.
- The recommendation report was basically a desktop study; no actual first-hand feedback was sought from end users.

The committee could not find any record of who finally took the decision to go for the wet-type screw compressor!

In the subject project, the compatibility of the gas with the lube oil had been duly checked by the contractor/compressor manufacturer as per stipulations in the contract.

Nobody could explain the phenomenon of the chock material in the lube oil.

The breakthrough came when one of the members of the committee had a chat with an operator who was an old-timer in the existing reciprocating compressor unit. He casually mentioned that they have been draining buckets of crude oil from the suction scrubber upstream of the reciprocating compressor, on a regular basis! Further, he noted that, the crude oil was getting even into the compressor cylinder as evidenced by the thick black coating in the valve plates of the compressors. He had seen it with his own eyes when the compressor was opened for maintenance.

He showed the committee member the half-barrel drum kept for draining the suction scrubber, which was full of crude oil! It was evident that the gas contained crude oil.

Based on the records and physical evidence from the site, the committee put together what could have gone wrong. The key points are as follows:

1. There was carryover of crude oil from the cone roof storage tank into the gas line. It was in the form of fine mist particles. On entering the suction scrubber, larger particles settled in the scrubber vessel, while the finer droplets went thru the mist eliminator in the suction scrubber and ended up in the lube oil, where it coagulated into the gluey material that chocked the filter.

2. The gas samples were taken from the main gas pipe using a small branch line. It would not have been a representative sample. Secondly, the analysis was done on dry gas. A better sampling technique would have been using an isokinetic setup, which was time-consuming and costly. Moreover, the carryover of crude droplets happens in a random style, which cannot be predicted easily, making a representative sample difficult.

3. While making the compressor selection and the report, the owner company and the FEED contractor did not get any first-hand feedback on performance of wet gas compressors, specifically in hydrocarbon gas service, from end users.

4. After some effort, the committee finally contacted another major oil company, only to find out that they also faced similar problems with the wet gas screw compressors and are somehow managing with additional filters and an intensive maintenance regime.

5. Contractually, the owner could not take over the units as the guarantee runs had failed.

It so happened that the compressor skid was assembled by a package vendor. The compressor alone was from the machine manufacturer, and the other components in the skid were designed by the EPC contractor and supplied by various vendors. It was difficult to pinpoint the responsibility of how the wrong compressor type was chosen. In summary, the compressor was not a fit to the system.

The issue ballooned up into a contractual and legal wrangle between the owner company and the EPC contractor.

The owner company (under the not-so-good advice of their contractual and legal teams) took a position that the gas analysis itself, particularly the C7+ component, implied that there were oil droplets in the incoming gas. And that the EPC contractor should have taken the above into account while designing the compressor skid. The technical teams were of the opinion that with certain modifications, the units could be made to work.

The EPC contractor's stand was that the owner's specifications did not specifically mention any crude carryover or liquid droplets in the gas. What was given was just the composition of the gas, in the gaseous state. The composition was presented in a usual data format, nothing unusual about it. If the voluminous FEED technical documentation contained all detailed data and information, why such important data on crude carryover as liquid mist were not stated clearly rather than in an implied manner!

The owner's senior management considered the opinions of all sides. They were inclined to go with the technical team's recommendation but for the thorny issue of sharing of the cost and accountability for project non-performance and delays that either side was not willing to concede. The contractual and legal aspects prevailed.

If one steps back and reviews the situation, actually both sides' objectives are the same, namely, that of completing the project and getting it up running. Unfortunately, the above perspective is lost when both sides try to prove each other wrong contractually and legally. In fact, in the end, both sides lost having spent a lot of time and money on litigation.

Analysis: The following factors are the key elements in what went wrong:

1. The type of compressor wet crew selected for the service is questionable. The wet type would have worked if the gas did not contain entrained liquid droplets (the liquid state of the micron-level droplets is important). The owner and the then FEED (engineering) contractor failed to understand the existing system properly and as a result selected a compressor that was a common choice in the industry, but for other services. As stated earlier, no first-hand end user experience was sought directly. Dry-type screw compressor would have worked in the situation.

2. The owner company's review of the recommendation of the FEED contractor was inadequate. First of all, when such studies are contracted out, there must be a clear understanding and responsibility for the FEED contractor on the deliverables and recommendations. Such desktop studies on equipment selection should be always supplemented with actual end user data.

   A FEED done under a tight schedule and focused deliverables is not a research project. Any uncertainties in the technical data or specifications should be studied separately and carried out by the right agencies. If required, pilot plant studies also have to be conducted. Owner companies should not shy away from such studies, and the time taken for the same.

   It is better for a project to fail in a pilot plant stage rather than on a full-scale project!

3. Further, conditional clauses in the EPC contract are more problematic. For example, in the subject project, there was the clause that the EPC contractor should verify the compatibility of the gas with the lube oil in the wet screw compressor. But it did not state what to do if the gas is not compatible with the lube oil!

   For completeness of the study, the proposal by the technical team for modification of the compressor unit is given below:

   The function of inlet gas scrubbers upstream of the compressor is to capture and take out any liquids in the gas. Conventional gas

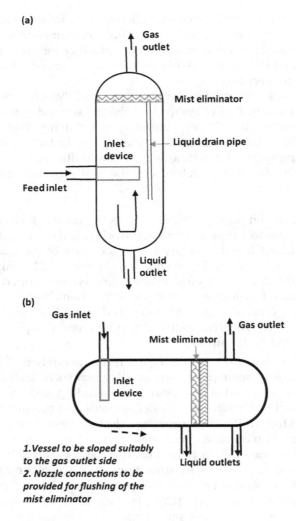

FIGURE 11.1
(a) Vertical inlet gas scrubber. (b) Horizontal inlet gas scrubber.

scrubbers can be vertical or horizontal. In the subject project, they had gone for a vertical vessel with a half-pipe inlet device and mesh time mist eliminator.

The technical team was of the opinion that given the fact of liquid carryover from the storage tank, a horizontal scrubber would have done the job better. The horizontal vessel has a better design, inlet device (preferably, tangential entry), and mist eliminator system consisting of vane- and mesh-type decks. The horizontal vessel can be sloped slightly for better liquid collection (see Figures 11.1a and b).

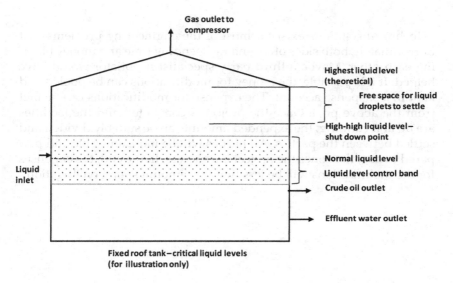

Fixed roof tank – critical liquid levels
(for illustration only)

FIGURE 11.2
Storage tank liquid levels.

The technical team with the help of the concerned operation personnel studied the entire system in detail. They found out that the high-high level in the storage tank is really high. The freeboard provided above the high-high level is not as per industry practice and not enough. The freeboard in this particular case should be higher considering the bubbling and agitated level inside the tank. This will be quite violent if some high gas wells get connected. The above situation is a recipe for carryover and should be recognized as such. Lowering the high-high level should help (see Figure 11.2).

KEY TAKEAWAYS

The above case study illustrates the importance of inlet FEED data and that it should not be treated lightly. Sufficient attention should be given to the same. The operations team can decide which data to be given, but a review by design personnel on the data will be helpful. Also, it is better to get user experience on equipment that looks suspicious for a particular service. Further, FEED should not have gray areas and conditional clauses.

Technical solutions exist for most of the engineering problems. But unfortunately, both sides often end up spending time and money blaming each other. Maybe a third-party specialist negotiator would have helped. If it is possible, the money for modifications can be pooled and kept in a suspense account. The expense for modifications can be met from the above pool. Once the project is completed and the facilities are up and running, the expended amount can be suitably divided and settled between the parties. All parties will be more amicable and prepared to negotiate once the project is a success. In the end, the expenses for the modifications will be less than the amount spend on litigation.

12

Iron Sulfide and Project Failure

Project overview: The following pages describe a project that failed to achieve its objectives due to unforeseen Iron Sulfide precipitation on the equipment. The importance of FEED composition is highlighted here.

Name: Effluent Treatment Project PQR

Client (owner company): Major oil company OWN5

FEED & Project Management Contractor: International engineering company (ENG5)

Contractor: International EPC contractor (EPC5)

Project location: Asia

Project duration: Varies from 24 to 30 months

Project value: About 500 million USD

Summary of project scope and description: EPC contractor scope included detailed engineering, procurement, installation, testing, commissioning, and handover of facilities to the owner. The project scope was to process a large volume of separated effluent water coming out from various oil and gas processing facilities.

Bidding and contract award: The Invitation to Bid (ITB) package with FEED technical documentation was prepared by an international engineering company (ENG5). There was a good response from international EPC contractors.

After several rounds of bid clarifications, the bid was awarded to EPC contractor EPC5, whose bid is accepted as technically suitable and the lowest.

What went wrong: The project was completed with minor delays. When the EPC contractor commenced commissioning, it was hampered after a few days due to choking of the media (Nutshell type) filter. On dismantling of the filter, it was found that the media was choked with a black precipitate, which was found to be iron sulfide later. The media was cleaned and put back. The thinking was that it may have precipitated due to chemical injections upstream of the filter. The chemical injection skids' dosage rates were adjusted, and the plant started up again.

However the problem persisted. Because of the large throughput involved, the quantity of the iron sulfide precipitation was also high, necessitating frequent shutdowns and cleaning of the media filters. This was happening despite the provision for a standby filter.

After several trials, the contractor recognized the situation as a major problem with inlet streams and formally wrote to the company about their inability to commission the plant. They squarely placed the problem on the

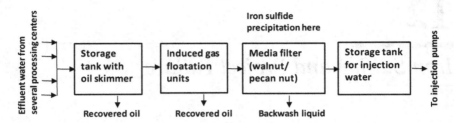

FIGURE 12.1
Effluent water treatment schematic.

inlet feed streams in front of the owner company and therefore requested the owner to handle the same.

Just to recap, the effluent water from the oil/gas/water separators, desalters, and storage tanks undergoes a series of processing steps as shown in Figure 12.1. The processing steps' main objective is to remove the oil and suspended solid particulate matter in the effluent water to the required specifications suitable for injection back into reservoirs for pressure maintenance or disposal. The processing steps involve a series of storage tanks and specialized vendor–designed equipment. During the processing steps, the effluent water quality improves, and the concentration of oil content and particulates decreases to target values, typically in the range of 20–30 parts per million (ppm) for oil and 95% removal of solids of 5 micron (1 micron$=10^{-6}$ m).

The project's large throughput could not be achieved due to the above issues. Over a period of running the plant, the contractor reduced the flow to about 60% of the original design, and managed the nutshell filter's backwash and cleaning cycles to keep the plant running. Guarantee runs were out of question.

Both the contractor and the owner engaged the services of separate specialized chemical companies to study the problem in detail and provide recommendations. It was a costly business involving 3 to 4 months' time including sample collections, site surveys, testing the samples, mixing studies of the effluent water from different fields in laboratories.

The reports and recommendations came eventually. The reports highlighted the following:

1. The formation of iron sulfide is not an unknown phenomenon in the oil and gas industry. The project management contractor's personnel were unaware of the importance of mixing, simply because they were design personnel, and did not have much idea about formation water and its properties. Furthermore, it was the first time that the owner company was undertaking a project where effluent water from several facilities was getting mixed.

2. Iron sulfide formation mechanism and chemistry are known. Iron sulfide in the form of FeS_2 generally precipitates out from the effluent water in two situations:

a. When water containing dissolved hydrogen sulfide comes into contact with equipment fabricated from carbon steel or iron materials.

b. When mixing water containing iron cations (Fe2) with another water containing hydrogen sulfide.

3. The first scenario was ruled out in the subject project. The second scenario could be causing the precipitation, though not in a simple manner on mixing alone, but through several complex mechanisms including the shearing of the oil droplets and agitation in the induced gas floatation units.

4. The temporary solution suggested was a modification of the existing chemical injection regime and the addition of a ferric sulfide dissolver chemical. It so happens that the ferric sulfide dissolver chemical is costly and would be a significant addition to the operating cost.

The above explanation is in layman's terms without going into details given in the voluminous technical reports produced by the consulting chemical companies.

With no other choice, the chemical injection regime was modified, and ferric sulfide dissolver chemical skids were added. Still, the capacity could not come up to the design values.

KEY TAKEAWAYS

It is ironic that, in retrospect, the company spent time and money on studying precisely what it did not do before or during the FEED. It could be that the project was rushed through conceptual and FEED stages, even though certain personnel may have wanted the specialized studies to be done. Using the typical jargon of 'fast tracking' the FEED will not help. In fact, pressurizing the FEED and shortening the FEED time is a fatal error an owner company can make in major projects. A quality FEED requires time – like most of the things in the world.

When the owner company's engineering staff itself is small, they depend on outsourced resources, who themselves may be totally unaware of the problems of oil field effluent water processing. Such knowledge could be available inside the owner company, but to utilize it requires very effective stakeholder management on the part of the project management team.

13

Provisions for Future and Associated Pitfalls

Project overview: The project was installation of a second train of effluent water processing facility adjacent to the existing first train. The owner company had provided adequate space for the new second train and also had made provisions in certain common facilities anticipating the new train at a later date. The project included detailed engineering, procurement, construction, and installation of equipment and associated utilities for an effluent water processing plant (second train).

Name: Effluent Water Treatment Project DSF

Client (owner company): Major oil company OWN6

FEED & Project Management Contractor: International engineering company (ENG6)

Contractor: International EPC contractor (EPC6)

Project location: Asia

Project duration: Varies from 30 to 36 months

Project value: About 500 million USD

Summary of project scope & description: EPC contractor scope included detailed engineering, procurement, installation, testing, commissioning, and handover of facilities to the owner. The project scope was to process a large volume of separated effluent water coming out from various oil and gas processing facilities. The scope specifically requested the contractor to utilize the space already provided in the existing main pipe rack in the first train to lay a large 24-inch diameter pipe, after checking the adequacy of the structural support and foundation design.

Bidding and contract award: The Invitation to Bid (ITB) package with FEED technical documentation was prepared by an international engineering company (ENG6). There was a good response from international EPC contractors.

After several rounds of bid clarifications, the bid was awarded to EPC contractor EPC6, whose bid is accepted as technically suitable and the lowest.

What went wrong: The project was going on smoothly when it hit a hurdle. The contractor did the checking after obtaining the design of the existing structural supports and foundations, which itself took some time. On completing the check in the civil engineering structural design software, the result came out stating that the existing pipe supports are not adequate to take the load of the new 24-inch pipe that is going to carry the effluent water. The space was adequate, though. A point worth noting is that the old-design

DOI: 10.1201/9781003317081-16

documentation for the first train did not contain any detail as to how and where the future provisions were given.

The suggestions from the contractor called for providing additional structural members throughout the entire length of the pipe rack, which was a long distance. Naturally, it involved additional cost and time from the contractual point of view, since the contract was not clear about what the contractor should do in case the adequacy verification came out the wrong way!

The project was stuck, and to resolve the issue, the owner company appointed a third party, a structural design expert to verify the contractor's calculations.

The structural design expert reviewed the contractor's calculations independently. They recommended that not only the structure needs to be strengthened, but the foundations also need strengthening!

The project was delayed due the contractual issues and the need to negotiate with the contractor to finalize the additional work involved. The cost of the additional work was not that much, but from a project point of view, a high-value project got delayed due to lack of diligence on the part of the designer of the first train.

KEY TAKEAWAYS

Provisions for future hardware and software can be made in a project, but the owner company needs to be extra careful while doing so. Certain importance points from the author (GUK) experience are noted as follows:

1. First of all, when such a design adequacy check is included in the scope of another EPC contractor, the FEED team needs to do a check by themselves first and take care to include the additional requirements. The EPC contractor check should be only to ensure fulfillment of the EPC contractor's responsibilities.

2. The FEED team while including design adequacy checks by the bidder/contractor in the ITB scope should ensure that the existing design documentation is available. Because the new facilities will be installed after several years, the accessibility of the 'as-built' documentation will be a question mark in most owner companies. To retrieve the old-design documentation after many years is not an easy task in many companies. Therefore, whenever clauses are put in the ITB for contractor verification of old designs, it presents a gray area. The owner and the FEED contractor need to be extra vigilant on such clauses.

3. It is suggested that in such cases where the company expects future facilities to come, a document titled 'Design Basis for Future Provisions' be prepared. The above document shall include the reasons and the objectives for providing provisions for future, as well as how it has been provided along with the list of documents where it is covered. With respect to software, the issue becomes more complicated due to version changes and compatibility issues. The author has seen facility DCS software being upgraded in one project, only to be dismantled and replaced with new software due to successive changes, additions, and modifications due to the lack of foresight and planning. The document mentioned above will be a big help when the company looks at the options for a second train 8 to 10 years from the construction of the first train.

4. Development of the FEED is one thing, but the financial and management issues and approvals are another. A FEED prepared can undergo changes when it goes through the financial and management approval. Such things happen due the project's cost estimate vs company budget constraints, time schedules that do not match with the overall development and production plans, etc. Further, most importantly, people and organizations change often resulting in shelving old and implementing new ideas. Managing change is still a difficult challenge in major projects despite the plethora of literature and tools available on the subject. Proper depository of information and its retrieval is easily possible with digital tools currently available, which should improve the decision-making during changes in FEED.

14

Heat Exchanger Selection and a Project Almost Gone Wrong

Project overview: The project scope consisted of adding new desalter and dehydrating trains in several oil processing centers. All trains were identical. The project included detailed engineering, procurement, construction, and installation of equipment and associated utilities. The outlet crude oil had to meet stringent quality with respect to salt and water content.

Name: Capacity Expansion Project ABC

Client (Owner company): Major oil company OWN7

FEED & Project Management Contractor: Inhouse and International engineering company (ENG7)

Contractor: International EPC contractor (EPC7)

Project location: Middle East

Project duration: Varies from 24 to 30 months

Project value: About 80 to 100 million USD

Summary of project scope & description: EPC contractor scope included detailed engineering, procurement, installation, testing, commissioning, and handover of desalter and dehydration trains to the owner. The project aimed to install an identical desalter and a dehydration train in several oil processing centers. Since the design was identical, any error in design had an impact on all the trains.

Bidding and contract award: The Invitation to Bid (ITB) package with FEED technical documentation was prepared by an international engineering company (ENG7).

After several rounds of bid clarifications, the bid was awarded to EPC contractor EPC7, whose bid is accepted as technically suitable and the lowest.

What almost went wrong: A typical desalter and dehydration train had several equipment types connected in series as shown in Figure 14.1. The heat input in the bath-type heater is optimized by transferring heat to cold inlet crude stream from hot outlet crude stream, thereby heating the cold crude prior to entry into the heater. The above heat exchange reduced the quantity of fuel gas required in the bath-type heater for heating up the crude oil.

Since the project was a major expansion of the facilities, there was a committee of experts to finalize the project scope and its key technical parameters. The committee had members from stakeholders from within the company, as well as external design consultants.

DOI: 10.1201/9781003317081-17

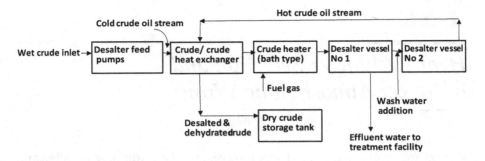

FIGURE 14.1
Schematic of the desalter train.

The committee finalized the Statement of Requirements (SOR) after several rounds of consultations with stakeholders, and the same was forwarded to the inhouse design and engineering unit of the company for preparation of the FEED.

The design started with the process simulation and preparation of process flow diagrams (PFDs). Since the process involved only liquid streams, the simulations were generally straightforward. Moreover, all the equipment types were also specified by the expert committee.

It so happened that the process simulation showed non-convergence of solution for one equipment, namely, the crude/crude heat exchanger, which was specified as shell & tube type. The error diagnostics of the simulation program noted 'temperature cross' as the cause for non-convergence.

The process design engineers could not really understand the situation initially. After doing some research, they understood the phenomenon as shown in highly simplified Figures 14.2a and b.

Without going into the heat exchanger design details, in summary, it makes sense that the outlet temperature of the heated cold stream cannot go above that of the outlet temperature of the cooled hot stream. Temperature cross means that the above principle has been violated and it is not physically possible to achieve the heat transfer required in one shell & tube exchanger. Thus, when the diagnostic said 'temperature cross', it meant that the specified one-number shell & tube exchanger cannot do the job!

There is one more term associated with the shell & tube exchanger design with counter-current hot and cold stream arrangement called 'temperature approach'. The temperature approach is the difference in the end point temperatures in an exchanger. In other words, it is a measure of how near the heated cold stream can approach the inlet hot stream as shown in Figure 14.2a. In fact, in shell & tube heat exchanger, for one unit, the temperature approach can be in the range of 10–20°C. As the temperature approach values come down, the feasibility of a single shell & tube achieving the required temperatures becomes doubtful.

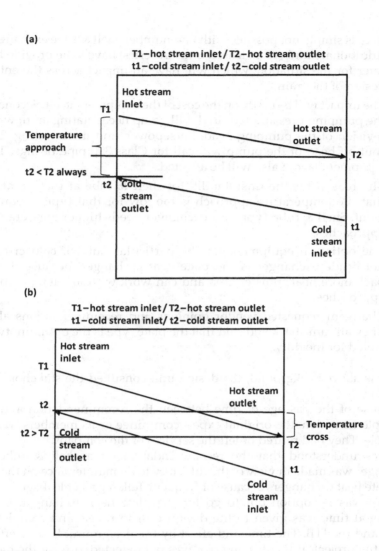

FIGURE 14.2
(a) Simplified diagram – normal temperature profile in a counter-current shell & tube heat exchanger. (b) Temperature profile with temperature cross.

The design engineers in the inhouse design unit did a detailed heat exchanger design. The results showed that:

1. The shell & tube heat exchanger as specified in the SOR document from the expert committee is not feasible with a standard-size single unit. The heat transfer required for the particular heat exchange

duty is simply not possible with one-number shell & tube exchanger. Additional numbers (at least 3 units in series) have to be provided to cater for the situation, which will have an impact across the entire design of the train.

2. The impact will be firstly on the cost of the equipment itself. Secondly, the pumping pressure required will go up necessitating an upward revision of the pumping head, horsepower, and motor rating. The shut-off head of the pump will call for Class 300 pipe fittings. The equipment layout also will be affected.

3. The reason for the unsuitability of shell & tube at that point, is that the temperature approach is too low for that type of equipment. Shell & tube-type heat exchanger needs higher temperature approaches.

4. The optimum equipment for the particular duty of cold crude/hot crude exchanger is the plate heat exchanger, because it can pack more heat transfer area and can work with low-temperature approaches.

5. The team requested vendor opinions and recommendations also. They all came back and said that the plate type is the optimum type suited for the duty.

With the above background, the design unit consulted the stakeholders on the issue.

Because of the passage of time between the recommendation and project implementation, the original expert committee or its members were not available. They had retired or left the services of the company.

It was understood that the recommendation on a shell & tube heat exchanger was made in view of the difficulty to do maintenance on the existing plate heat exchangers because of frequent Teflon gasket leakage.

There was no option but to go for the plate heat exchanger. Certain additional time was given to the design unit to revise and complete the FEED and the ITB. The time and effort by the design unit had its effect. It saved the project. If the design error was not detected, during the bidding process some of the EPC contractors may have found this out and challenged the design, probably resulting in cancellation of the bid, with dire consequences.

If it was not detected during the bidding, the consequences would have been worse and difficult to handle contractually, since the EPC contractor would have requested for a massive change order.

It was found that the frequent gasket leakage in existing plate heat exchangers of the owner company, was due to incorrect assembly of the gasket, the plate and the frame!. The original equipment manufacturer (OEM), was glad to train the owners maintenance personnel for the above work.

KEY TAKEAWAYS

A conceptual process design is a key element that is often overlooked in the early stages of the project. It requires a certain way of thinking and willingness to explore alternatives in design solutions. Equipment type and sizes have an immediate impact on the equipment layout of the facility and cascading impacts on other disciplines. The engineering manager's role in the above aspects is crucial.

The importance of setting up and running process simulation models at the earliest opportunity in a project is emphasized. Original Equipment Manufacturers (OEMs) must be consulted when choices of equipment look suspicious. Changes to existing established systems should be studied in depth before modifications are brought in.

15

Too Many Unknowns – FEED Gone Wrong and Its Impacts

Project overview: This case study is based on an international offshore EPC project executed for a project in South-East Asia. The project is unique in many aspects, mainly because of the entities involved and the uncertainties under which the project was tendered out. The terminology used here is different from the other case studies.

The three primary parties involved in this project are as follows:

1. A South Asian oil and gas exploration and production company (the owner or client).
2. A South-East Asian country who has granted concession to the owner/client for gas exploration in their territorial waters (the partner country).
3. A South-East Asian EPC contractor who was awarded the contract for the project (the contractor).

Name: Wellhead platforms WHP1, WHP2, and WHP3, and connected pipelines.

FEED: The client provided FEED in the ITB, which formed the technical basis of the contractor's bid and subsequently served as a technical basis of the EPC contract.

The key issues that dominated the proposal stage are stated as follows:

1. The bid mandated performance of *FEED Verification and Endorsement*.
2. The FEED containing Design Dossier included various high-end engineering analyses such as Spectral Fatigue Analysis, Transportation Fatigue Analysis, and Floatation & Upending Analysis, and other stringent technical specifications.
3. The high-end engineering analysis required highly competent engineers with relevant experience, specialized software, and adequate time, which were constraints.

Project location: South-East Asia
  Project duration: 14 months
  Project value: Lump sum 325 million USD

DOI: 10.1201/9781003317081-18

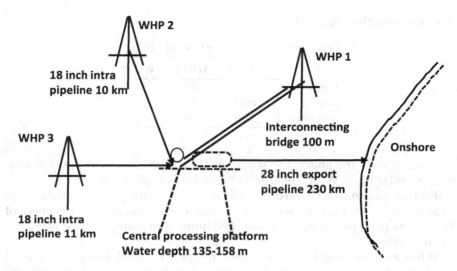

FIGURE 15.1
Project scope: WHP and pipelines.

Summary of project scope & description: The project scope consisted of three offshore wellhead platforms (WHP1, WHP2, and WHP3) and pipelines connecting them to a Central Processing Platform (being installed by another contractor). A part of the gas production was to be taken off for domestic use in the partner country, and the remaining gas was for export by the client.

The scope of work included detailed engineering, procurement, supply, construction, fabrication, transportation, installation, offshore hook-up, pre-commissioning, and commissioning of three wellhead platforms and the connecting pipelines.

The primary steel for structures and coated/lined pipes for pipelines were supplied by the client as a free-issue material.

The scope is visualized in Figure 15.1. Dotted lines are not in the scope of the subject project.

Unique Features of the Project: Several aspects of this project were unique with many firsts for all the three principal participants:

1. The first offshore project in this area of the sea.

2. Extremely high water depth of 158 meters, deepest so far for the client and the contractor.

3. Single-piece piles of length up to 144 meters.

4. Installation of long slender jacket by the launch method and long pile installation by the pile upending method.

5. Rare Intensity Earthquake Analysis for a return period of 2500 years.

Key Commodity Quantities

Component	Quantity (Steel) in Tons		
	WHP1	WHP2	WHP3
Topside	1225	1450	1450
Jacket	6925	8100	7100
Skirt piles	4225	2800	3850

In addition to the above, the scope included two intra-field sealines connecting WHP2 and WHP3 platforms to the process platform.

Bidding and contract award: The EPC contract for the project was awarded in May 2012, with the time schedule for handover of the plant to the client of 14 months from the contract award. The lump sum contract price was USD 325 million.

What went wrong: From the outset, the project faced a large number of implementation challenges starting from the engineering stage, through procurement and fabrication, right up to the installation stage. A summary of the challenges faced is the following:

1. A large number of very specialized structural and safety engineering analysis mandated in the contract required the contractor to mobilize multiple specialist agencies. The results of such analysis resulted in major changes to the original structural design considered by the contractor at the proposal stage, with a considerable impact on project cost and schedule.

2. Many vendors around whose equipment the FEED was structured abstained from participating during project execution, because the partner country came under a trade embargo from US and European countries. The search for alternate vendors, the process of getting the client approval for new vendors, and the platform layout changes resulting from the changes in equipment design put huge pressure on cost and schedule.

3. The selection and finalization of marine assets such as transportation barges and installation barges ran into big headwinds due to the limited availability of such assets and simultaneous implementation of a large number of offshore projects all around the world. Here again, the result was high pressure on the cost and schedule.

4. At the installation stage, it was found that the piling strategy was not working due to an incorrect soil analysis report, which was included in the FEED. This error was discovered only when initial piling attempts hit an unforeseen hard stratum, resulting in pile refusal. Root cause analysis, alternate piling design, and sourcing of

additional equipment and resources, all resulted in additional costs and time. At this stage, time delay was a huge cost due to marine asset deployment schedule extension, resulting in large penalties.

Contractual Mitigation Measures by the Contractor: The contractor's project control team had been proactively keeping track of any event, incident, or condition that led to changes in terms of escalation of activity duration and costs to the original plan. To deliver a focused approach to Change management, workshops were arranged for engineering and fabrication teams to make them aware and sensitize them to variations occurring and need to be vigilant in the identification of changes. The workshops facilitated open discussion between teams to identify additional change orders and record it in a Change register.

Apart from contracts team, members from various disciplines took additional responsibility for providing discipline-specific leads and supported the contracts team in formulation and timely submission of change orders to clients. All the notifications of the change orders were issued within 10 days of the incidence, and the Change Request was compiled and submitted to the client within the stipulated period of 15 days.

The end result of this diligent approach was that before the installation phase was completed, the contractor had progressively placed before the client multiple change orders totaling USD 60 million in cost and about 30 days for an extension of the time schedule.

Negotiation and Settlement of Claims: The final negotiation between the client and the contractor was scheduled when the fabrication of WHP1 was close to completion and the sail away date of the WHP1 deck was fast-approaching. There were substantial disagreements on the validity of certain change orders and the extension of time (EOT) claim already submitted. After protracted discussions with the client, the contractor made strategic moves to persuade the client to redefine the terms of the contract to address all pending claims and linked it to sail away of the WHP1 deck as per an accelerated schedule.

The client and the contractor finally reached an agreement with USD 38 million in additional costs and a 25-day extension to the project completion schedule.

The Project Completion Status: The project was completed within the extended project schedule, but the contractor suffered a cost overrun of USD 47 million (USD 35 million loss of profit + USD 12 million net loss over the enhanced contract price). The contractor decided to accept this loss as there were some errors on their part in the proposal stage engineering and estimation, primarily associated with the marine asset mobilization and platform fabrication subcontracts.

KEY TAKEAWAYS

1. The FEED for offshore projects must be specific and conclusive with respect to all issues that can substantially impact the cost or implementation schedule of the project. Typically, these should include all specialized structural and safety studies, and soil analysis for the installation location. No contractor would be able to carry out these activities during the pre-bid period due to time and resource constraints.

2. The FEED should not be overly reliant on equipment or packages of specific vendors, because the availability of such vendors at the project implementation stage can be uncertain due to a full workload, a closure of business, or political issues between countries. Getting alternate vendors onboard requires substantial time for contractors and clients and could significantly affect design aspects such as layout and foundation loading.

3. Political issues like a trade embargo impacting the availability of suppliers and subcontractors to support a project in a country/region are an issue in 'no man's land' and need a negotiated resolution between contractor and client.

4. For offshore projects, a key aspect is the availability and cost of marine assets, both of which fluctuate wildly depending on the number of offshore projects under implementation at the same time around the world. Attractive costs at the bidding stage could turn prohibitive by the time the implementation is taken up. Long-term relationships and a good overview of the global project scenario are essential to ensure reasonable compliance between estimation and implementation.

16

When FEED Is Uncertain, Contracting Strategy Can Help

Without diminishing the importance and value of a comprehensive and accurate FEED, we need to recognize that there would be instances when generating such a FEED becomes very difficult for the owner. Such projects by themselves could be very critical for the owner, and rather than abandoning them due to the difficulty in producing a reliable FEED, a way must be found to implement such a project with a minimum risk to the owner, as well as the project implementing agency. There are several examples worldwide of how such projects are successfully implemented, and given below is a case study of one such project.

Project Overview: A national oil company operating in Asia has operating assets established over a period from mid-1950 to late 1960. The production from these assets is critical not just for the company, but also for the country, whose economy is heavily dependent on this production. These assets have undergone running repairs and periodic modifications through their operating life. The assets are spread across the country as discrete processing units, and this particular project deals with the following types of units:

1. Oil gathering units, which receive crude oil from oil wells located nearby, process the crude oil to remove water and associated gas, and store and export the treated crude oil to an export terminal. The associated gas is compressed and sent to gas processing stations for cleaning, condensate recovery, and finally exportation to power generation plants in the country. The recovered water is sent to water injection facilities, which inject the water back into the oil wells.

2. Gas processing units, which receive the associated gas from the oil gathering units, remove water from the gas, compress the gas, recover condensate, and sent the dry compressed gas to power stations.

Over the years, the owner's internal safety audits, as well as the audits carried out by their insurers, had identified some very critical areas of safety risk, and at a point in time, it was deemed imperative that these safety risks have to be mitigated. The critical safety risks were as follows:

1. The operating units were very old and had been subject to repairs and modifications carried out without proper documentation or records. The result was that there was no engineering documentation

DOI: 10.1201/9781003317081-19

available with the owner as to the process flows, control systems, electrical power distribution systems, and the overall unit layout.

2. The entire interconnecting piping between the equipment, and the instrumentation and electrical cabling were routed underground, and no routing diagram or layout was available. The buried pipes and cables were deemed as a big safety risk because their condition could not be periodically checked for corrosion, leakage, insulation deterioration, etc.

Project location: Asia
Project duration: 24 to 36 month phased completion
Project value: About 2 billion USD
Summary of project scope and description: The owner decides to go ahead with a mitigation project and ropes in a project management contractor (PMC) to assist with formulating a project scope and specification documents.

The scope included certain new equipment/systems to improve the operational efficiency, and combined with the basic safety mitigation measures, the overall project scope was as follows:

1. New desalters and three-phase separators at the oil gathering units.
2. New condensate recovery compressor systems in the oil gathering units.
3. New process control systems in both the oil gathering and gas processing units.
4. New utility systems for instrument air, nitrogen, and electric power receiving and distribution to cater to the added load of the new equipment/systems.
5. Relocation of all underground piping and cabling to new aboveground pipe/cable racks.
6. Generate all critical 'as-built' documentation covering all the existing and new equipment, systems, and facilities.

For items 1 to 4 above, the owner's engineering team had already put together a FEED.

Project Contracting Strategy: The owner and the PMC reviewed the total project scope and the FEED. They concluded that a clear and definitive FEED is available for the entire scope barring one area, which is the relocation of all underground piping and cabling to new aboveground pipe/cable racks, using new pipes and cables. This is a gray area because there is no existing documentation available as to the number of pipes and cables, which are currently buried underground, and also, there is no clarity possible on the route of the new aboveground pipe/cable racks, because they have to be routed to avoid interference with the existing buried pipes and cables.

The owner and PMC decide on a contracting strategy, which will provide an optimum value for the owner, as well as provide a level playing field for an EPC contractor to make a fair assessment of his costs and risks. The key points of the overall contracting strategy are as follows:

1. A fixed lump sum contract for a detailed survey of the various units to be modified to create an 'as is' documentation for process flow, instrument and control systems, electrical power receiving and distribution systems, and existing equipment layout. This survey was also intended to identify location of all the new equipment, and the routing of new pipe/cable racks without interference with the existing underground pipes and cables.

2. A fixed lump sum contract for the new equipment and systems to be installed, which can be clearly defined with scope and specifications.

3. A re-measurable unit rate contract for building the new aboveground pipe/cable racks and laying the new pipes and cables thereon.

4. The entire project construction is to be done in a 'brownfield' environment, which means that all the units will be working as normal, while the new construction and installation works take place within the Units. There will be a very specific Unit shutdown window varying between one week and three weeks in each Unit to facilitate the interconnection of all the new equipment, control systems, power receiving and distribution systems, pipes, and cables, to the existing equipment, after cutting out the existing pipes and cables. This was a very critical contract requirement because this determines the extent of production shortfalls for oil and gas the company has to plan for, which in turn will impact the country's oil export commitments, as well as domestic power generation targets.

The project, due to its size and complexity, was awarded to multiple EPC contractors, with clearly defined Units to be modified by each contractor. The overall value of this project was in excess of USD 2 billion, with a progressive and phased completion timeframe between 24 and 36 months.

Project Implementation: The projects were implemented successfully by the EPC contractors. More than 40 million manhours were spent by EPC contractors in project implementation, the predominant portion of which was used for brownfield construction work. Despite this large volume of brownfield work, there were only a few minor incidences of 'lost time' on the project and the project had a zero-fatality record. All the EPC contractors earned a decent profit on the project, and the owner had more efficient and safer operation facilities completed with a minimum loss of production.

KEY TAKEAWAYS

1. A contracting strategy that requires a fixed lump sum commitment for the areas, which are clearly defined and specified, and allows a re-measurable value for the areas, which cannot be clearly defined is a 'win-win' situation for the owner, as well as the contractors.

2. Two key requirements of the owner on this project were the absolute safety during the brownfield construction phase, which had potentially huge risks to owner's assets and personnel, and adherence to the contract specified shutdown windows in each Unit, as this was linked to the national oil export commitments and domestic power generation targets. The contracting strategy adopted by the owner recognized that putting unnecessary cost pressure on contractors could jeopardize these key requirements and that a fair contracting strategy would avoid such unnecessary cost pressures on the contractors and allow them to focus on the key client requirements.

Bibliography

A Guide to Project Management Body of Knowledge (PMBOK Guide) (6th ed.). (2017). Project Management Institute.

Asmar, M.E., & Gibson, G.E. Jr. (2018). The cost impact of Front End Engineering Design (FEED) accuracy for large industrial projects. In Abdul-Malak, M., Khoury, H., Singh, A., & Yazdani, S. (Eds.). *Responsible Design and Delivery of the Constructed Project.* ISEC Press, Fargo, ND, USA. Vol. 5(1). ISSN: 2644-108X, ISBN: 978-0-9960437-5-5. Retrieved Sep 26, 2023, from https://www.isec-society.org/ISEC_PRESS/EURO_MED_SEC_02/pdf/CPM-06.pdf

Baron, H. (2016). Why our way to develop projects fails to minimize CAPEX-Part 1. *Journal of Petroleum Technology.* Retrieved Sep 26, 2023, from https://jpt.spe.org/why-our-way-develop-projects-fails-minimize-capexpart-1

Baron, H. (2017). Why our way to develop projects fails to minimize CAPEX-Part 2. *Journal of Petroleum Technology.* Retrieved Sep 26, 2023, from https://jpt.spe.org/why-our-way-develop-projectsfails-minimize-capexpart-2

Barshop, P. (2009). Incomplete FEL 2-destroyer of capital. *Independent Project Analysis Newsletter.* Independent Project Analysis, Inc., Ashburn. Retrieved Feb 28, 2023, from https://www.ipaglobal.com/wp-content/uploads/2019/01/IPA-Newsletter-2009-Q4-Volume-1-Issue-4.pdf

Cabano, S., Balcezak, M., Gibson, E. Jr., West, M., Garison, G., & Ochsner, E. (2017, Jul 31–Aug 2). *Front End Engineering Design Maturity and Accuracy Total Rating System* [Conference presentation]. CII Annual Conference, Orlando, FL, USA. Retrieved Feb 28, 2023, from https://www.construction-institute.org/CII/media/Publications/presentations/2017/RT-331_slides.pdf

Concept of Novation in EPC (Engineering, Procurement & Construction) Industry's Contracts. The Next Level Purchasing Association (NLPA). North Carolina, USA. Retrieved Sep 26, 2023, from https://www.certitrek.com/nlpa/blog/concept-of-novation-in-epc-engineering-procurement-construction-industrys-contracts/

Construction Industry Institute. (2018). *IR331-2 - Front End Engineering Design (FEED) Maturity and Accuracy Total Rating System (MATRS).* The University of Texas at Austin.

Haider, T. (2013). The board room. 4 out of every 5 oil gas megaprojects fail. But why. Interview with Edward Merrow. Oil and gas IQ. Retrieved Feb 28, 2023, from https://www.oilandgasiq.com/oil-and-gas-production-and-operations/interviews/interview-4-out-of-every-5-oil-gas-megaprojects-fail

Landry, L. (2022). Why emotional intelligence is important in leadership. Retrieved Feb 28, 2023, from https://online.hbs.edu/blog/post/emotional-interlligence-in-leadership

Littau, P., Dunović, I.B., Pau, L-F., Mancini, M., Dieguez, A.I., Medina-Lopez, C., et al. (2015). Managing stakeholders in megaprojects-the MS working group report. *European Cooperation in Science & Technology (COST).* University of Leeds. Retrieved Sep 26, 2023, from https://bib.irb.hr/datoteka/768614.Managing_Stakeholders_in_Megaprojects.pdf

Menon, S.E. (2015). *Pipeline Calculations and Simulations Manual*. Elsevier.

Merrow, E.W. (2011). *Industrial Megaprojects Concepts, Strategies, and Practices for Success*. Wiley.

Merrow, E.N., & Nandurdikar, N. (2018). *Leading Complex Projects: A Data-Driven Approach to Mastering the Human Side of Project Management*. Wiley.

Rawlins, C.H., & Sadeghi, F. (2017). Experimental study on oil removal in nutshell filters for produced-water treatment. *SPE Production & Operations* 33 (01): 145–153. Paper Number: SPE-186104-PA. Retrieved Sep 23, 2023, from https://www.semanticscholar.org/paper/Experimental-Study-on-Oil-Removal-in-Nutshell-for-Rawlins-Sadeghi/a1634f5b1569246e21c36eff77b7e06daf486db8

Shlopak, M., Emblemsvåg, J., & Oterhals, O. (2014). Front end loading as an integral part of the project execution model in lean shipbuilding. *Proceedings IGLC-22*, (pp. 207–220), Oslo, Norway. Retrieved Sep 26, 2023, from https://iglcstorage.blob.core.windows.net/papers/attachment-018e8ada-ba8c-43be-9031-cd4f80341a59.pdf

Swanson, T., & Laperouse, D. (2019). Next-generation system mitigates iron sulfide. *Hart Enery Newsletter*. Retrieved Sep 26, 2023, from https://www.hartenergy.com/exclusives/next-generation-system-mitigates-iron-sulfide-182315

Index

Note: **Bold** page numbers refer to tables, *italic* page numbers refer to figures.

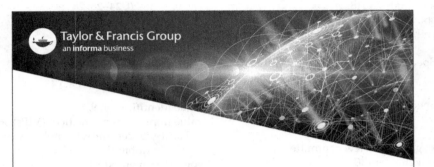

Printed in the United States
by Baker & Taylor Publisher Services